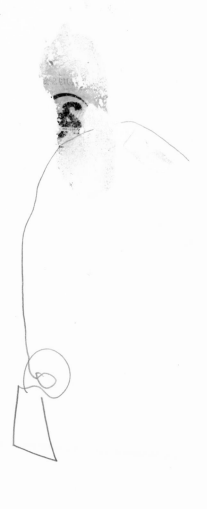

Origins

Also by Neil deGrasse Tyson

The Sky Is Not the Limit: Adventures of an Urban Astrophysicist

Cosmic Horizons: Astronomy at the Cutting Edge
(with Steven Soter, eds.)

One Universe: At Home in the Cosmos (with Charles Liu and
Robert Irion)

Universe Down to Earth

Just Visiting This Planet

Merlin's Tour of the Universe

Also by Donald Goldsmith

Chaos to Cosmos: A Space Odyssey
(with Laura Danly and Leonard David)

*Connecting with the Cosmos: Nine Ways to Experience
the Wonder of the Universe*

The Search for Life in the Universe (with Tobias Owen; 3rd ed.)

*The Runaway Universe: The Race to Find
the Future of the Cosmos*

The Ultimate Planets Book

Worlds Unnumbered: The Search for Extrasolar Planets

The Ultimate Einstein (with Robert Libbon)

*Einstein's Greatest Blunder? The Cosmological Constant and
Other Fudge Factors in the Physics of the Universe*

Origins

Fourteen Billion Years of Cosmic Evolution

Neil deGrasse Tyson

and

Donald Goldsmith

W. W. NORTON & COMPANY

NEW YORK · LONDON

For information about permission to reproduce selections from this book, write to Permissions, W. W. Norton & Company, Inc., 500 Fifth Avenue, New York, NY 10110

Manufacturing by R. R. Donnelley, Harrisonburg
Book design by Chris Welch
Production manager: Amanda Morrison

Library of Congress Cataloging-in-Publication Data

Tyson, Neil deGrasse.
Origins : fourteen billion years of cosmic evolution / Neil deGrasse Tyson
and Donald Goldsmith.— 1st ed.
p. cm.
Includes bibliographical references and index.
ISBN 0-393-05992-8
1. Cosmology. 2. Evolution. 3. Life—Origin. I. Goldsmith, Donald. II. Title.
QB981.T96 2004
523.1—dc22

2004012201

W. W. Norton & Company, Inc., 500 Fifth Avenue, New York, N.Y. 10110
www.wwnorton.com

W. W. Norton & Company Ltd., Castle House, 75/76 Wells Street, London W1T 3QT

1 2 3 4 5 6 7 8 9 0

To all those who look up,
And to all those who do not yet know
why they should

Contents

9

Part III: The Origin of Stars

Part IV: The Origin of Planets

Part V: The Origin of Life

Acknowledgments

For reading and rereading the manuscript, ensuring that we mean what we say and say what we mean, we are indebted to Robert Lupton of Princeton University. His tandem expertise in astrophysics and the English language allowed the book to reach several notches higher than we had otherwise imagined for it. We are also grateful to Sean Carroll at Chicago's Fermi Institute, Tobias Owen of the University of Hawaii, Steven Soter of the American Museum of Natural History, Larry Squire of UC San Diego, Michael Strauss of Princeton University, and PBS NOVA producer Tom Levenson for key suggestions that improved several parts of the book.

For expressing confidence in the project from the beginning, we thank Betsy Lerner of the Gernert Agency, who saw our manuscript not only as a book but also as an expression of deep interest in the cosmos, deserving the broadest possible audience with whom to share the love.

Major portions of Part II and scattered portions of Parts I and

III first appeared as essays in *Natural History* magazine by NDT. For this, he is grateful to Peter Brown, the magazine's editor in chief, and especially to Avis Lang, their senior editor, who continues to work heroically as a learned literary shepherd to NDT's writing efforts.

The authors further recognize support from the Sloan Foundation in the writing and preparation of this book. We continue to admire their legacy of support for projects such as this.

Neil deGrasse Tyson, New York City

Donald Goldsmith, Berkeley, California

June 2004

Origins

A Meditation on the Origins of Science and the Science of Origins

Anew synthesis of scientific knowledge has emerged and continues to flourish. In recent years, the answers to questions about our cosmic origins have not come solely from the domain of astrophysics. Working under the umbrella of emergent fields with names such as astrochemistry, astrobiology, and astro-particle physics, astrophysicists have recognized that they can benefit greatly from the collaborative infusion of other sciences. To invoke multiple branches of science when answering the question, Where did we come from? empowers investigators with a previously unimagined breadth and depth of insight into how the universe works.

In *Origins: Fourteen Billion Years of Cosmic Evolution*, we introduce the reader to this new synthesis of knowledge, which allows us to address not only the origin of the universe but also the origin of the largest structures that matter has formed, the origin of the stars that light the cosmos, the origin of planets that offer the likeliest sites for life, and the origin of life itself on one or more of those planets.

Humans remain fascinated with the topic of origins for many reasons, both logical and emotional. We can hardly comprehend the essence of anything without knowing where it came from. And of all the stories that we hear, those that recount our own origins engender the deepest resonance within us.

Self-centeredness bred into our bones by our evolution and experience on Earth has led us naturally to focus on local events and phenomena in the retelling of most origin stories. However, every advance in our knowledge of the cosmos has revealed that we live on a cosmic speck of dust, orbiting a mediocre star in the far suburbs of a common sort of galaxy, among a hundred billion galaxies in the universe. The news of our cosmic unimportance triggers impressive defense mechanisms in the human psyche. Many of us unwittingly resemble the man in the cartoon who gazes at the starry heavens and remarks to his companion, "When I look at all those stars, I'm struck by how insignificant they are."

Throughout history, different cultures have produced creation myths that explain our origins as the result of cosmic forces shaping our destiny. These histories have helped us to ward off feelings of insignificance. Although origin stories typically begin with the big picture, they get down to Earth with impressive speed, zipping past the creation of the universe, of all its contents, and of life on Earth, to arrive at long explanations of myriad details of human history and its social conflicts, as if we somehow formed the center of creation.

Almost all the disparate answers to the quest of origins accept as their underlying premise that the cosmos behaves in accordance with general rules, which reveal themselves, at least in principle, to our careful examination of the world around us. Ancient Greek philosophers raised this premise to exalted heights, insisting that we humans possess the power to perceive how nature operates, as well as the underlying reality beneath what we observe: the fundamental truths that govern all else. Quite understandably, they insisted that uncovering those truths would be difficult. Twenty-three hundred years ago, in his most famous reflection on our ignorance, the Greek philosopher Plato compared those who strive for knowledge to prisoners chained in a cave, unable to see objects behind them, and who must attempt to deduce from the shadows of these objects an accurate description of reality.

With this simile, Plato not only summarized humanity's attempts to understand the cosmos but also emphasized that we have a natural tendency to believe that mysterious, dimly sensed entities govern the universe, privy to knowledge that we can, at best, glimpse only in part. From Plato to Buddha, from Moses to Mohammed, from a hypothesized cosmic creator to modern films about "the matrix," humans in every culture have concluded that higher powers rule the cosmos, gifted with an understanding of the gulf between reality and superficial appearance.

Half a millennium ago, a new approach toward understanding nature slowly took hold. This attitude, which we now call science, arose from the confluence of new technologies and the discoveries that they fostered. The spread of printed books across Europe, together with simultaneous improvements in travel by road and water, allowed individuals to communicate more quickly and effectively, so that they could learn what others had to say and could respond far more rapidly than in the past. During the sixteenth and seventeenth centuries, this hastened back-and-forth disputation and led to a new way of acquiring knowledge, based on the principle that the most effective means of understanding the cosmos relies on careful observations, coupled with attempts to specify broad and basic principles that explain a set of these observations.

One more concept gave birth to science. Science depends on organized skepticism, that is, on continual, methodical doubting.

Few of us doubt our own conclusions, so science embraces its skeptical approach by rewarding those who doubt someone else's. We may rightly call this approach unnatural; not so much because it calls for mistrusting someone else's thoughts, but because science encourages and rewards those who can demonstrate that another scientist's conclusions are just plain wrong. To other scientists, the scientist who corrects a colleague's error, or cites good reasons for seriously doubting his or her conclusions, performs a noble deed, like a Zen master who boxes the ears of a novice straying from the meditative path, although scientists correct one another more as equals than as master and student. By rewarding a scientist who spots another's errors—a task that human nature makes much easier than discerning one's own mistakes—scientists as a group have created an inborn system of self-correction. Scientists have collectively created our most efficient and effective tool for analyzing nature, because they seek to disprove other scientists' theories even as they support their earnest attempts to advance human knowledge. Science thus amounts to a collective pursuit, but a mutual admiration society it is not, nor was meant to be.

Like all attempts at human progress, the scientific approach works better in theory than in practice. Not all scientists doubt one another as effectively as they should. The need to impress scientists who occupy powerful positions, and who are sometimes swayed by factors that lie beyond their conscious knowledge, can interfere with science's self-correcting ability. In the long run, however, errors cannot endure, because other scientists will discover them and promote their own careers by trumpeting the news. Those conclusions that do survive the attacks of other scientists will eventually achieve the status of scientific "laws," accepted as valid descriptions of reality, even though scientists understand that each of these laws may some day find itself to be only part of a larger, deeper truth.

But scientists hardly spend all their time attempting to prove one another mistaken. Most scientific endeavors proceed by testing imperfectly established hypotheses against slightly improved observational results. Every once in a while, however, a significantly new take on an important theory emerges, or (more often in an age of technological advances) a whole new range of observations opens the way to a new set of hypotheses to explain these new results. The greatest moments in scientific history have arisen, and will always arise, when a new explanation, perhaps coupled with new observational results, produces a seismic shift in our conclusions about the workings of nature. Scientific progress depends on individuals in both camps: those who assemble better data and extrapolate carefully from it; and those who risk much—and have much to gain if successful—by challenging widely accepted conclusions.

Science's skeptical core makes it a poor competitor for human hearts and minds, which recoil from its ongoing controversies and prefer the security of seemingly eternal truths. If the scientific approach were just one more interpretation of the cosmos, it would never have amounted to much; but science's big-time success rests on the fact that it works. If you board an aircraft built according to science—with principles that have survived numerous attempts to prove them wrong—you have a far better chance of reaching your destination than you do in an aircraft constructed by the rules of Vedic astrology.

Throughout relatively recent history, people confronted with the success of science in explaining natural phenomena have reacted in one of four ways. First, a small minority have embraced the scientific method as our best hope for understanding nature, and seek no additional ways to comprehend the universe. Second, a much larger number ignore science, judging it uninteresting, opaque, or opposed to the human spirit. (Those who watch television greedily without ever pausing to wonder where the pictures and sound come from remind us that the words "magic" and "machine" share deep etymological roots.) Third, another minority, conscious of the assault that science seems to make upon their cherished beliefs, seek actively to disprove scientific results that annoy or enrage them. They do so, however, quite outside the skeptical framework of science, as you can easily establish by asking one of them, "What evidence would convince you that you are wrong?" These anti-scientists still feel the shock that John Donne described in his poem "The Anatomy of the World: The First Anniversary," written in 1611 as the first fruits of modern science appeared:

And new philosophy calls all in doubt,
The element of fire is quite put out,
The Sun is lost, and th'earth, and no man's wit
Can well direct him where to look for it.
And freely men confess that this world's spent,
When in the planets and the firmament
They seek so many new; they see that this [world]
Is crumbled out again to his atomies.
'Tis all in pieces, all coherence gone . . .

Fourth, another large section of the public accepts the scientific approach to nature while maintaining a belief in supernatural entities existing beyond our complete understanding that rule the cosmos. Baruch Spinoza, the philosopher who created the strongest bridge between the natural and the supernatural, rejected any distinction between nature *and* God, insisting instead that the cosmos is simultaneously nature and God. Adherents of more conventional religions, which typically insist on this distinction, often reconcile the two by mentally separating the domains in which the natural and the supernatural operate.

No matter what camp you may live in, no one doubts that these are auspicious times for learning what's new in the cosmos. Let us

then proceed with our adventurous quest for cosmic origins, acting much like detectives who deduce the facts of the crime from
the evidence left behind. We invite you to join us in search of cosmic clues—and the means of interpreting them—so that
together we may uncover the story of how part of the universe
turned into ourselves.

Overture

The Greatest Story Ever Told

> The world has persisted many a long year, having once been set going in the appropriate motions. From these everything else follows.
>
> —Lucretius

Some 14 billion years ago, at the beginning of time, all the space and all the matter and all the energy of the known universe fit within a pinhead. The universe was then so hot that the basic forces of nature, which collectively describe the universe, were merged into a single, unified force. When the universe was a roaring 10^{30} degrees and just 10^{-43} seconds old—the time before which all of our theories of matter and space lose their meaning—black holes spontaneously formed, disappeared, and formed again out of the energy contained within the unified force field. Under these extreme conditions, in what is admittedly speculative physics, the structure of space and time became severely curved as it gurgled into a spongy, foamlike structure. During this epoch, phenomena described by Einstein's general theory of relativity (the modern theory of gravity) and quantum mechanics (the description of matter on its smallest scales) were indistinguishable.

As the universe expanded and cooled, gravity split from the

other forces. Soon thereafter, the strong nuclear force and the electro-weak force split from each other, an event accompanied by an enormous release of stored energy that induced a rapid, fifty-power-of-ten increase in the size of the universe. The rapid expansion, known as the "epoch of inflation," stretched and smoothed matter and energy so that any variation in density from one part of the universe to the next became less than one part in a hundred thousand.

Continuing onward with what is now laboratory-confirmed physics, the universe was hot enough for photons to spontaneously convert their energy into matter-antimatter particle pairs, which immediately thereafter annihilated each other, returning their energy back to photons. For reasons unknown, this symmetry between matter and antimatter had been "broken" at the previous force splitting, which led to a slight excess of matter over antimatter. The asymmetry was small but crucial for the future evolution of the universe: for every 1 billion antimatter particles, 1 billion+1 matter particles were born.

As the universe continued to cool, the electro-weak force split into the electromagnetic force and the weak nuclear force, completing the four distinct and familiar forces of nature. While the energy of the photon bath continued to drop, pairs of matter-antimatter particles could no longer be created spontaneously from the available photons. All remaining pairs of matter-antimatter particles swiftly annihilated, leaving behind a universe with one particle of ordinary matter for every billion photons—and no antimatter. Had this matter-over-antimatter asymmetry not emerged, the expanding universe would forever be composed of light and nothing else, not even astrophysicists. Over a roughly three-minute period, the matter became protons and neutrons, many of which combined to become the simplest atomic nuclei. Meanwhile, free-roving electrons thoroughly scattered the photons to and fro, creating an opaque soup of matter and energy.

When the universe cooled below a few thousand degrees Kelvin —somewhat hotter than a blast furnace—the loose electrons moved slowly enough to get snatched from the soup by the roving nuclei to make complete atoms of hydrogen, helium, and lithium, the three lightest elements. The universe had now become (for the first time) transparent to visible light, and these free-flying photons are observable today as the cosmic microwave background. During its first billion years, the universe continued to expand and cool as matter gravitated into the massive concentrations we call galaxies. Within just the volume of the cosmos that we can see, a hundred billion of these galaxies formed, each containing hundreds of billions of stars that undergo thermonuclear fusion in their cores. Those stars with more than about ten times the mass of the Sun achieve sufficient pressure and temperature in their cores to manufacture dozens of elements heavier than hydrogen, including the elements that compose planets and the life upon them. These elements would be embarrassingly useless were they to remain locked inside the star. But high-mass stars explode in death, scattering their chemically enriched guts throughout the galaxy.

After 7 or 8 billion years of such enrichment, an undistinguished star (the Sun) was born in an undistinguished region (the Orion arm) of an undistinguished galaxy (the Milky Way) in an undistinguished part of the universe (the outskirts of the Virgo supercluster). The gas cloud from which the Sun formed contained a sufficient supply of heavy elements to spawn a few planets, thousands of asteroids, and billions of comets. During the formation of this star system, matter condensed and accreted out of the parent cloud of gas while circling the Sun. For several hundred million years, the persistent impacts of high-velocity comets and other leftover debris rendered molten the surfaces of the rocky planets, preventing the formation of complex molecules. As less and less accretable matter remained in the solar system, the planets' surfaces began to cool. The planet we call Earth formed in an orbit where its atmosphere can sustain oceans, largely in liquid form. Had Earth formed much closer to the Sun, the oceans would have vaporized. Had Earth formed much farther, the oceans would have frozen. In either case, life as we know it would not have evolved.

Within the chemically rich liquid oceans, by a mechanism unknown, simple anaerobic bacteria emerged that unwittingly transformed Earth's carbon dioxide–rich atmosphere into one with sufficient oxygen to allow aerobic organisms to form, evolve, and dominate the oceans and land. These same oxygen atoms, normally found in pairs (O_2), also combined in threes to form ozone (O_3) in the upper atmosphere, which shields Earth's surface from most of the Sun's molecule-hostile ultraviolet photons.

The remarkable diversity of life on Earth, and (we may presume) elsewhere in the universe, arises from the cosmic abundance of carbon and the countless number of molecules (simple and complex) made from it; more varieties of carbon-based molecules exist than of all other molecules combined. But life is fragile. Earth's encounters with large objects, left over from the formation of the solar system, which were once common events, still wreak intermittent havoc upon our ecosystem. A mere 65 million years ago (less than 2 percent of Earth's past), a 10-trillion-ton asteroid struck what is now the Yucatán Peninsula and obliterated over 70 percent of Earth's land-based flora and fauna-including all the dinosaurs, the dominant land animals of that epoch. This ecological tragedy opened an opportunity for small, surviving mammals to fill freshly vacant niches. A big-brained branch of these mammals, one we call primates, evolved a genus and species—*Homo sapiens*—to a level of intelligence that enabled them to invent methods and tools of science; to invent astrophysics; and to deduce the origin and evolution of the universe.

Yes, the universe had a beginning. Yes, the universe continues to evolve. And yes, every one of our body's atoms is traceable to the big bang and to the thermonuclear furnaces within high-mass stars. We are not simply in the universe, we are part of it. We are born from it. One might even say that the universe has empowered us, here in our small corner of the cosmos, to figure itself out. And we have only just begun.

Part I

The Origin of the Universe

CHAPTER 1

In the Beginning

n the beginning, there was physics. "Physics" describes how matter, energy, space, and time behave and interact with one another. The interplay of these characters in our cosmic drama underlies all biological and chemical phenomena. Hence everything fundamental and familiar to us earthlings begins with, and rests upon, the laws of physics. When we apply these laws to astronomical settings, we deal with physics writ large, which we call astrophysics.

In almost any area of scientific inquiry, but especially in physics, the frontier of discovery lives at the extremes of our ability to measure events and situations. In an extreme of matter, such as the neighborhood of a black hole, gravity strongly warps the surrounding space-time continuum. At an extreme of energy, thermonuclear fusion sustains itself within the 15-million-degree cores of stars. And at every extreme imaginable we find the outrageously hot and dense conditions that prevailed during the first few moments of the universe. To understand what happens in each of these scenarios requires laws of physics discovered after

1900, during what physicists now call the modern era, to distinguish it from the classical era that includes all previous physics.

One major feature of classical physics is that events and laws and predictions actually make sense when you stop and think about them. They were all discovered and tested in ordinary laboratories in ordinary buildings. The laws of gravity and motion, of electricity and magnetism, and of the nature and behavior of heat energy are still taught in high school physics classes. These revelations about the natural world fueled the industrial revolution, itself transforming culture and society in ways unimagined by generations that came before, and remain central to what happens, and why, in the world of everyday experience.

By contrast, nothing makes sense in modern physics because everything happens in regimes that lie far beyond those to which our human senses respond. This is a good thing. We may happily report that our daily lives remain wholly devoid of extreme physics. On a normal morning, you get out of bed, wander around the house, eat something, then dash out the front door. At day's end your loved ones fully expect you to look no different than you did when you left, and to return home in one piece. But imagine yourself arriving at the office, walking into an overheated conference room for an important 10 A.M. meeting, and suddenly losing all your electrons—or worse yet, having every atom of your body fly apart. That would be bad. Suppose instead that you're sitting in your office trying to get some work done by the light of your 75-watt desk lamp, when somebody flicks on 500 watts of overhead lights, causing your body to bounce randomly from wall to wall until you're jack-in-the-boxed out the window. Or what if you go to a sumo wrestling match after work, only to see the two nearly spherical gentlemen collide, disappear, and then spontaneously become two beams of light that leave the room in opposite directions? Or suppose that on your way home, you take a road less traveled, and a darkened building sucks you in feet first, stretching your body head to toe while squeezing you shoulder to shoulder as you get extruded through a hole, never to be seen or heard from again.

If those scenes played themselves out in our daily lives, we would find modern physics far less bizarre; our knowledge of the foundations of relativity and quantum mechanics would flow naturally from our life experiences; and our loved ones would probably never let us go to work. But back in the early minutes of the universe that kind of stuff happened all the time. To envision it, and to understand it, we have no choice but to establish a new form of common sense, an altered intuition about how matter behaves, and how physical laws describe its behavior, at extremes of temperature, density, and pressure.

We must enter the world of $E = mc^2$.

Albert Einstein first published a version of this famous equation in 1905, the year in which his seminal research paper entitled "Zur Elektrodynamik bewegter Körper" appeared in *Annalen der Physik*, the preeminent German journal of physics. The paper's title in English reads "On the Electrodynamics of Moving Bodies," but the work is far better known as Einstein's special theory of relativity, which introduced concepts that forever changed our notions of space and time. Just twenty-six years old in 1905, working as a patent examiner in Bern, Switzerland, Einstein offered further details, including his best-known equation in another, remarkably short (two-and-a-half-page) paper published later the same year in the same journal: "Ist die Trägheit eines Körpers von seinem Energieinhalt abhängig?" or "Does the Inertia of a Body Depend on Its Energy Content?" To save you the effort of locating the original article, of designing an experiment, and of thus testing Einstein's theory, the answer to the paper's title is yes. As Einstein wrote,

If a body gives off the energy E in the form of radiation, its mass diminishes by E/c². . . .The mass of a body is a measure of its energy-content; if the energy changes by E, the mass changes in the same sense.

Uncertain as to the truth of his statement, he then suggested,

It is not impossible that with bodies whose energy-content is variable to a high degree (e.g. with radium salts) the theory may be successfully put to the test.[*]

There you have it: the algebraic recipe for all occasions when you want to convert matter into energy, or energy into matter. $E = mc^2$—energy equals mass times the square of the speed of light—gives us a supremely powerful computational tool that extends our capacity to know and understand the universe from as it is now, all the way back to infinitesimal fractions of a second after the birth of the cosmos. With this equation, you can tell how much radiant energy a star can produce, or how much you could gain by converting the coins in your pocket into useful forms of energy.

The most familiar form of energy—shining all around us, though often unrecognized and unnamed in our mind's eye—is the photon, a massless, irreducible particle of visible light, or of any other form of electromagnetic radiation. We all live within a continuous bath of photons: from the Sun, the Moon, and the stars; from your stove, your chandelier, and your nightlight; from hundreds of radio and television stations; and from countless cellphone and radar transmissions. Why, then, don't we actually see the daily transmuting of energy into matter, or of matter into energy? The energy of common photons sits far below the mass of the least massive subatomic particles, when converted into energy

[*] Albert Einstein, *The Principle of Relativity*, trans. by W. Perrett and G. B. Jeffery (London: Methuen and Company, 1923), pp. 69–71.

by $E = mc^2$. Because these photons wield too little energy to become anything else, they lead simple, relatively uneventful lives.

Do you long for some action with $E = mc^2$? Start hanging around gamma-ray photons that have some real energy—at least 200,000 times more than visible photons. You'll quickly get sick and die of cancer; but before that happens, you'll see pairs of electrons, one made of matter, the other of antimatter (just one of many dynamic particle-antiparticle duos in the universe) pop into existence where photons once roamed. As you watch, you'll also see matter-antimatter pairs of electrons collide, annihilating each other and creating gamma-ray photons once again. Increase the photons' energy by another factor of 2,000, and you now have gamma rays with enough energy to turn susceptible people into the Hulk. Pairs of these photons wield enough energy, fully described by the power of $E = mc^2$, to create particles such as neutrons, protons, and their antimatter partners, each nearly 2,000 times the mass of an electron. High-energy photons don't hang out just anywhere, but they do exist in many a cosmic crucible. For gamma rays, almost any environment hotter than a few billion degrees will do just fine.

The cosmological significance of particles and energy packets that transform themselves into one another is staggering. Currently, the temperature of our expanding universe, found by measuring the bath of microwave photons that pervades all of space, is a mere 2.73 degrees Kelvin. (On the Kelvin scale, all temperatures are positive: particles have the least possible energy at 0 degrees; room temperature is about 295 degrees; and water boils at 373 degrees.) Like the photons of visible light, microwave photons are too cool to have any realistic ambitions of turning themselves into particles via $E = mc^2$. In other words, no known particle has a mass so low that it can be made from the meager energy of a microwave photon. The same holds true for the photons that form radio waves, infrared, and visible light, as well as ultraviolet and X rays. More simply expressed, particle transmutations all require gamma rays. Yesterday, however, the universe was a little bit smaller and a little bit hotter than today. The day before, it was smaller and hotter still. Roll the clocks backward some more—say, 13.7 billion years—and you land squarely in the post–big bang primordial soup, a time when the temperature of the cosmos was high enough to be astrophysically interesting as gamma rays filled the universe.

To understand the behavior of space, time, matter, and energy from the big bang to present day is one of the greatest triumphs of human thought. If you seek a complete explanation for the events of the earliest moments, when the universe was smaller and hotter than ever thereafter, you must find a way to enable the four known forces of nature—gravity, electromagnetism, the strong and the weak nuclear forces—to talk to one another, to unify and become a single meta-force. You must also find a way to reconcile two currently incompatible branches of physics: quantum mechanics (the science of the small) and general relativity (the science of the large).

Spurred by the successful marriage of quantum mechanics and electromagnetism during the mid-twentieth century, physicists moved swiftly to blend quantum mechanics and general relativity into a single and coherent theory of quantum gravity. Although so far they have all failed, we already know where the high hurdles lie: during the "Planck era." That's the cosmic phase up to 10^{-43} second (one ten-million-trillion-trillion-trillionth of a second) after the beginning. Because information can never travel more rapidly than the speed of light, 3×10^8 meters per second, a hypothetical observer situated anywhere in the universe during the Planck era could see no farther than 3×10^{-35} meter (three hun-

dred billion trillion-trillionths of a meter). The German physicist
Max Planck, after whom these unimaginably small times and distances are named, introduced the idea of quantized energy in
1900 and generally receives credit as the father of quantum
mechanics.

Not to worry, though, so far as daily life goes. The clash
between quantum mechanics and gravity poses no practical problem for the contemporary universe. Astrophysicists apply the
tenets and tools of general relativity and quantum mechanics to
extremely different classes of problems. But in the beginning,
during the Planck era, the large was small, so there must have
been a kind of shotgun wedding between the two. Alas, the vows
exchanged during that ceremony continue to elude us, so no
(known) laws of physics describe with any confidence how the
universe behaved during the brief honeymoon, before the
expanding universe forced the very large and very small to part
ways.

At the end of the Planck era, gravity wriggled itself loose from
the other, still-unified forces of nature, achieving an independent
identity nicely described by our current theories. As the universe
aged past 10^{-35} second, it continued to expand and to cool, and
what remained of the once-unified forces divided into the electro-
weak force and the strong nuclear force. Later still, the electro-
weak force split into the electromagnetic and the weak nuclear
forces, laying bare four distinct and familiar forces—with the
weak force controlling radioactive decay, the strong force binding
together the particles in each atomic nucleus, the electromagnetic
force holding atoms together in molecules, and gravity binding
matter in bulk. By the time the universe aged a trillionth of a second, its transmogrified forces, along with other critical episodes,
had already imbued the cosmos with its fundamental properties,
each worthy of its own book.

While time dragged on for the universe's first trillionth of a second, the interplay of matter and energy continued incessantly. Shortly before, during, and after the strong and electro-weak forces had split, the universe contained a seething ocean of quarks, leptons, and their antimatter siblings, along with bosons, the particles that enable these particles to interact with one another. None of these particle families, so far as we now know, can be divided into anything smaller or more basic. Fundamental though they are, each family of particles comes in several species. Photons, including those that form visible light, belong to the boson family. The leptons most familiar to the nonphysicist are electrons and (perhaps) neutrinos; and the most familiar quarks are . . . well, there are no familiar quarks, because in ordinary life we always find quarks bound together into particles such as protons and neutrons. Each species of quark has been assigned an abstract name that serves no real philological, philosophical, or pedagogical purpose except to distinguish it from the others: "up" and "down," "strange" and "charmed," and "top" and "bottom."

Bosons, by the way, derive their name from the Indian physicist Satyendranath Bose. The word "lepton" comes from the Greek *leptos*, meaning "light" or "small." "Quark," however, has a literary and far more imaginative origin. The American physicist Murray Gell-Mann, who in 1964 proposed the existence of quarks, and who then thought that the quark family had only three members, drew the name from a characteristically elusive line in James Joyce's *Finnegans Wake*: "Three quarks for Muster Mark!" One advantage quarks can claim: All their names are simple—something that chemists, biologists, and geologists seem unable to achieve in naming their own stuff.

Quarks are quirky. Unlike protons, which each have an electric charge of $+1$, and electrons, each with a charge of -1, quarks have fractional charges that come in units of $1/3$. And except under the most extreme conditions, you'll never catch a quark all by itself; it will always be clutching on to one or two other quarks. In fact, the force that keeps two (or more) of them together actually grows *stronger* as you separate them—as if some sort of subnuclear rubber band held them together. Separate the quarks sufficiently far, and the rubber band snaps. The energy stored in the stretched band now summons $E = mc^2$ to create a new quark at each end, leaving you back where you started.

During the quark-lepton era in the cosmos's first trillionth of a second, the universe had a density sufficient for the average separation between unattached quarks to rival the separation between attached quarks. Under those conditions, allegiances between adjacent quarks could not be established unambiguously, so they moved freely among themselves. The experimental detection of this state of matter, understandably named "quark soup," was reported for the first time in 2002 by a team of physicists working at the Brookhaven National Laboratories on Long Island.

The combination of observation and theory suggests that an episode in the very early universe, perhaps during one of the splits between different types of force, endowed the cosmos with a remarkable asymmetry, in which particles of matter outnumbered particles of antimatter by only about one part in a billion— a difference that allows us to exist today. That tiny discrepancy in population could hardly have been noticed amid the continuous creation, annihilation, and recreation of quarks and antiquarks, electrons and anti-electrons (better known as positrons), and neutrinos and antineutrinos. During that era, the odd man out—the slight preponderance of matter over antimatter—had plenty of opportunities to find other particles with which to annihilate, and so did all the other particles.

But not for much longer. As the universe continued to expand and cool, its temperature fell rapidly below 1 trillion degrees Kelvin. A millionth of a second had now passed since the beginning, but this tepid universe no longer had a temperature or density sufficient to cook quarks. All the quarks quickly grabbed dance partners, creating a permanent new family of heavy particles called hadrons (from the Greek *hadros*, meaning "thick"). That quark-to-hadron transition quickly produced protons and neutrons as well as other, less familiar types of heavy particles, all composed of various combinations of quarks. The slight matter-antimatter asymmetry in the quark-lepton soup now passed to the hadrons, with extraordinary consequences.

As the universe cooled, the amount of energy available for the spontaneous creation of particles declined continuously. During the hadron era, photons could no longer invoke $E = mc^2$ to manufacture quark-antiquark pairs: their E could not cover the pairs' mc^2. In addition, the photons that emerged from all the remaining annihilations continued to lose energy to the ever-expanding universe, so their energies eventually fell below the threshold required to create hadron-antihadron pairs. Every billion annihilations left a billion photons in their wake—and only a single hadron survived, mute testimony to the tiny excess of matter over antimatter in the early universe. Those lone hadrons would ultimately get to have all the fun that matter can enjoy: they would provide the source of galaxies, stars, planets, and people.

Without the imbalance of a billion and one to a mere billion between matter and antimatter particles, all the mass in the universe (except for the dark matter whose form remains unknown) would have annihilated before the universe's first second had passed, leaving a cosmos in which we could see (if we had existed) photons *and nothing else*—the ultimate Let-there-be-light scenario.

By now, one second of time has passed.

At 1 billion degrees, the universe remains piping hot—still able to cook electrons, which, along with their positron (antimatter) counterparts, continue to pop in and out of existence. But within the ever-expanding, ever-cooling universe, their days (seconds, really) are numbered. What was formerly true for hadrons now comes true for electrons and positrons: they annihilate each other, and only one electron in a billion emerges, the lone survivor of the matter-antimatter suicide pact. The other electrons and positrons died to flood the universe with a greater sea of photons.

With the era of electron-positron annihilation over, the cosmos has "frozen" into existence one electron for every proton. As the cosmos continues to cool, with its temperature falling below 100 million degrees, its protons fuse with other protons and with neutrons, forming atomic nuclei and hatching a universe in which 90 percent of these nuclei are hydrogen and 10 percent are helium, along with relatively tiny numbers of deuterium, tritium, and lithium nuclei.

Two minutes have now passed since the beginning.

Not for another 380,000 years does much happen to our particle soup of hydrogen nuclei, helium nuclei, electrons, and photons. Throughout these hundreds of millennia, the cosmic temperature remains sufficiently hot for the electrons to roam free among the photons, batting them to and fro.

As we will shortly detail in Chapter 3, this freedom comes to an abrupt end when the temperature of the universe falls below 3,000 degrees Kelvin (about half the temperature of the Sun's surface). Right about now, all the electrons acquire orbits around the nuclei, forming atoms. The marriage of electrons with nuclei leaves the newly formed atoms within a ubiquitous bath of visible light photons, completing the story of how particles and atoms formed in the primordial universe.

As the universe continues to expand, its photons continue to lose energy. Today, in every direction astrophysicists look, they find a cosmic fingerprint of microwave photons at a temperature of 2.73 degrees, which represents a thousandfold decline in the photons' energies since the time atoms first formed. The photons' patterns on the sky—the exact amount of energy that arrives from different directions—retain a memory of the cosmic distribution of matter just before atoms formed. From these patterns, astrophysicists can obtain remarkable knowledge, including the age and shape of the universe. Even though atoms now form part of daily life in the universe, Einstein's equation still has plenty of work to do—in particle accelerators, where matter-antimatter particle pairs are created routinely from energy fields; in the core of the Sun, where 4.4 million tons of matter are converted into energy every second; and in the cores of all other stars.

$E = mc^2$ also manages to apply itself near black holes, just outside their event horizons, where particle-antiparticle pairs can pop into existence at the expense of the black hole's formidable gravitational energy. The British cosmologist Stephen Hawking first described the hijinks in 1975, showing that the entire mass of a black hole can slowly evaporate by this mechanism. In other words, black holes are not entirely black. The phenomenon is known as Hawking radiation, and serves as a reminder of the continued fertility of Einstein's most famous equation.

But what happened *before* all this cosmic fury? What happened before the beginning?

Astrophysicists have no idea. Rather, our most creative ideas have little or no grounding in experimental science. Yet the religious faithful tend to assert, often with a tinge of smugness, that something must have started it all: a force greater than all others, a source from which everything issues. A prime mover. In the mind of such a person that something is, of course, God, whose nature varies from believer to believer but who always bears the responsibility for starting the ball rolling.

But what if the universe was always there, in a state or condition that we have yet to identify—a multiverse, for example, in which everything we call the universe amounts to only a tiny bubble in an ocean of suds? Or what if the universe, like its particles, just popped into existence from nothing we could see?

These rejoinders typically satisfy no one. Nonetheless, they remind us that informed ignorance provides the natural state of mind for research scientists at the ever-shifting frontiers of knowledge. People who believe themselves ignorant of nothing have neither looked for, nor stumbled upon, the boundary between what is known and unknown in the cosmos. And therein lies a fascinating dichotomy. "The universe always was," gets no respect as a legitimate answer to "What was around before the beginning?" But for many religious people, the answer, "God always was," is the obvious and pleasing answer to "What was around before God?"

No matter who you may be, engaging yourself in the quest to discover where and how everything began can induce emotional fervor—as if knowing our beginnings would bestow upon you some form of fellowship with, or perhaps governance over, all that comes later. So what is true for life itself is true for the universe: knowing where you came from is no less important than knowing where you are going.

CHAPTER 2

Antimatter Matters

Particle physicists have won the contest for the most peculiar, yet playful jargon of all the physical sciences. Where else could you find a neutral vector boson exchanged between a negative muon and a muon neutrino? Or a gluon exchange binding together a strange quark and a charmed quark? And where else can you meet squarks, photinos, and gravitinos? Alongside these seemingly countless particles with peculiar names, particle physicists must contend with a parallel universe of *anti*particles, collectively known as antimatter. Despite its persistent appearance in science fiction stories, antimatter is real. And as you might suppose, it does tend to annihilate upon contact with ordinary matter.

The universe reveals a peculiar romance between antiparticles and particles. They can be born together out of pure energy, and they can annihilate as they reconvert their combined mass back into energy. In 1932, the American physicist Carl David Anderson discovered the anti-electron, the positively charged, antimatter counterpart to the negatively charged electron. Since then, particle physicists have routinely made antiparticles of all varieties in the world's particle accelerators, but only recently have they assembled antiparticles into whole atoms. Since 1996, an international group led by Walter Oelert of the Institute for Nuclear Physics Research in Jülich, Germany, has created atoms of antihydrogen, in which an anti-electron happily orbits an antiproton. To make these first anti-atoms, the physicists used the giant particle accelerator operated by the European Organization for Nuclear Research (better known by its French acronym CERN) in Geneva, Switzerland, where so many important contributions to particle physics have occurred.

The physicists use a simple creation method: make a bunch of anti-electrons and a bunch of antiprotons, bring them together at a suitable temperature and density, and wait for them to combine to form atoms. During their first round of experiments, Oelert's team produced nine atoms of antihydrogen. But in a world dominated by ordinary matter, life as an antimatter atom can be precarious. The antihydrogen atoms survived for less than 40 nanoseconds (40 billionths of a second) before annihilating with ordinary atoms.

The discovery of the anti-electron was one of the great triumphs of theoretical physics, for its existence had been predicted just a few years earlier by the British-born physicist Paul A. M. Dirac.

To describe matter on the smallest size scales—those of atomic and subatomic particles—physicists developed a new branch of physics during the 1920s to explain the results of their experiments with these particles. Using newly established rules, now known as quantum theory, Dirac postulated from a second solution to his equation that a phantom electron from the "other side" might occasionally pop into the world as an ordinary electron, leaving behind a gap or hole in the sea of negative energies.

Although Dirac hoped to explain protons in this way, other physicists suggested that this hole would reveal itself experimentally as
a positively charged anti-electron, which had come to be known as
a positron for its positive electric charge. The detection of actual
positrons confirmed Dirac's basic insight and established antimatter as worthy of as much respect as matter.

Equations with double solutions are not unusual. One of the
simplest examples answers the question, What number times
itself equals nine? Is it 3 or −3? Of course, the answer is both,
because 3 x 3 = 9 and −3 x −3 = 9. Physicists cannot guarantee
that all the solutions of an equation correspond to events in the
real world, but if a mathematical model of a physical phenomenon is correct, manipulating its equations can be as useful as
(and somewhat easier than) manipulating the entire universe. As
with Dirac and antimatter, such steps often lead to verifiable predictions. If the predictions prove incorrect, then the theory is dis-
carded. But regardless of the physical outcome, a mathematical
model ensures that the conclusions you may draw from it are both
logical and internally consistent.

Subatomic particles have many measurable features, of which
mass and electric charge rank among the most important. Except
for the particle's mass, which is always the same for a particle and
its antiparticle, the specific properties of each type of antiparticle
will always be precisely opposite to those of the particle for which
it provides the "anti." For example, the positron has the same
mass as the electron, but the positron has one unit of positive
charge while the electron has one unit of negative charge. Similarly, the antiproton provides the oppositely charged antiparticle
of the proton.

Believe it or not, the chargeless neutron also has an antiparticle. It's called—you guessed it—the antineutron. An antineutron has an opposite zero charge with respect to the ordinary
neutron. This arithmetical magic derives from the particular

triplet of fractionally charged particles (the quarks) that form neutrons. The three quarks that compose a neutron have charges of $-\frac{1}{3}$, $-\frac{1}{3}$, and $+\frac{2}{3}$, while those in the antineutron have charges of $\frac{1}{3}$, $\frac{1}{3}$, and $-\frac{2}{3}$. Each set of three quark charges adds up to zero net charge, yet the corresponding components do have opposite charges.

Antimatter can pop into existence out of thin air. If gamma-ray photons have sufficiently high energy, they can transform themselves into electron-positron pairs, thus converting all of their seriously large energy into a small amount of matter, in a process whose energy side fulfills Einstein's famous equation $E = mc^2$.

In the language of Dirac's original interpretation, the gamma-ray photon kicked an electron out of the domain of negative energies, creating an ordinary electron and an electron hole. The reverse process can also occur. If a particle and an antiparticle collide, they will annihilate by refilling the hole and emitting gamma rays. Gamma rays are the sort of radiation you should avoid.

If you somehow manage to manufacture a blob of antiparticles at home, you have a wolf by the ears. Storage would immediately become a challenge, because your antiparticles would annihilate with any conventional sack or grocery bag (either paper or plastic) in which you chose to confine or carry them. A cleverer storage mechanism involves trapping the charged antiparticles within the confines of a strong magnetic field, where they are repelled by invisible but highly effective magnetic "walls." If you embed the magnetic field in a vacuum, you can render the antiparticles safe from annihilation with ordinary matter. This magnetic equivalent of a bottle will also be the bag of choice whenever you must handle other container-hostile materials, such as the 100-million-degree glowing gases involved in (controlled) nuclear fusion experiments. The greatest storage problem arises after you have created entire anti-atoms, because anti-atoms, like atoms, do not normally rebound from a magnetic wall. You would be wise to keep your positrons and antiprotons in separate magnetic bottles until you must bring them together.

To generate antimatter requires at least as much energy as you can recover when it annihilates with matter to become energy again. Unless you had a full tank of antimatter fuel before launch, a self-generating antimatter engine would slowly suck energy from your starship. Perhaps the original *Star Trek* television and film series embodied this fact, but if memory serves, Captain Kirk continually asked for "more power" from the matter-antimatter drives, and Scotty invariably replied in his Scottish accent that "the engines canna take it."

Although physicists expect hydrogen and antihydrogen atoms to behave identically, they have not yet verified this prediction experimentally, mainly because of the difficulty they face in keeping antihydrogen atoms in existence, rather than having them annihilate almost immediately with protons and electrons. They would like to verify that the detailed behavior of a positron bound to an antiproton in an antihydrogen atom obeys all the laws of quantum theory, and that an anti-atom's gravity behaves precisely as we expect of ordinary atoms. Could an anti-atom produce antigravity (repulsive) instead of ordinary gravity (attractive)? All theory points toward the latter, but the former, if it should prove correct, would offer amazing new insights into nature. On atomic-size scales, the force of gravity between any two particles is immeasurably small. Instead of gravity, electromagnetic and nuclear forces dominate the behavior of these tiny particles, because both forces are much, much stronger than gravity. To test for antigravity, you would need enough anti-atoms to make ordinary-sized objects, so that you can measure their bulk properties and compare them to ordinary matter. If a set of billiard balls (and, of course, the billiard table and the cue sticks) were made of antimatter, would a game of anti-pool be indistinguishable from a game of pool? Would an anti–eight ball fall into the corner pocket in exactly the same way as an ordinary eight ball? Would anti-planets orbit an anti-star the way that ordinary planets orbit ordinary stars?

It's philosophically sensible, and in line with all the predictions of modern physics, to presume that the bulk properties of antimatter will prove to be identical to those of ordinary matter—normal gravity, normal collisions, normal light, and so forth. Unfortunately, this means that if an anti-galaxy were headed our way, on a collision course with the Milky Way, it would remain indistinguishable from an ordinary galaxy until it was too late to do anything about it. But this fearsome fate cannot be common in the universe today because if, for example, a single anti-star annihilated with a single ordinary star, the conversion of their matter and antimatter into gamma-ray energy would be swift, violent, and total. If two stars with masses similar to the Sun's (each containing 10^{57} particles) were to collide in our galaxy, their melding would produce an object so luminous that it would temporarily outproduce all the energy of all the stars of 100 million galaxies and fry us to an untimely end. We have no compelling evidence that such an event has ever occurred anywhere in the universe. So, best we can judge, the universe is dominated by ordinary matter, and has been since the first few minutes after the big bang. Thus total annihilation through matter-antimatter collisions need not rank among your chief safety concerns on your next intergalactic voyage.

Still, the universe now seems disturbingly imbalanced: we expect particles and antiparticles to be created in equal numbers, yet we find a cosmos dominated by ordinary particles, which seem to be perfectly happy without their antiparticles. Do hidden pockets of antimatter in the universe account for the imbalance? Was a law of physics violated (or was an unknown law of physics at work?) during the early universe, forever tipping the balance in favor of matter over antimatter? We may never know the answers

to these questions, but for now, if an alien hovers over your front
lawn and extends an appendage as a gesture of greeting, toss it
your eight ball before you get too friendly. If the appendage and
the ball explode, the alien probably consists of antimatter. (How
it and its followers will react to this result, and what the explosion
will do to you, need not detain us here.) And if nothing untoward
occurs, you can proceed safely to take your new friend to your
leader.

CHAPTER 3

Let There Be Light

At the time when the universe was just a fraction of a second old, a ferocious trillion degrees hot, and aglow with an unimaginable brilliance, its main agenda was expansion. With every passing moment the universe got bigger as more space came into existence from nothing (not easy to imagine, but here, evidence speaks louder than common sense). As the universe expanded, it grew cooler and dimmer. For hundreds of millennia, matter and energy cohabited in a kind of thick soup in which speedy electrons continually scattered photons of light to and fro.

Back then, if your mission had been to see across the universe, you couldn't have done so. Any photons entering your eye would, just nanoseconds or picoseconds earlier, have bounced off electrons right in front of your face. You would have seen only a glowing fog in all directions, and your entire surroundings—luminous, translucent, reddish-white in color—would have been nearly as bright as the surface of the Sun.

As the universe expanded, the energy carried by each photon

decreased. Eventually, about the time that the young universe reached its 380,000th birthday, its temperature dropped below 3,000 degrees, with the result that protons and helium nuclei could permanently capture electrons, thus bringing atoms into the universe. In previous epochs, every photon had sufficient energy to break apart a newly formed atom, but now the photons had lost this ability, thanks to the cosmic expansion. With fewer unattached electrons to gum up the works, the photons could finally race through space without bumping into anything. That's when the universe became transparent, the fog lifted, and a cosmic background of visible light was set free.

That cosmic background persists to this day, the remnant of leftover light from a dazzling, sizzling early universe. It's a ubiquitous bath of photons, acting as much like waves as they do like particles. Each photon's wavelength equals the separation between one of its wiggly wave crests and the next—a distance you could measure with a ruler, if you could get your hands on a photon. All photons travel at the same speed in a vacuum, 186,000 miles per second (naturally called the speed of light), so photons with shorter wavelengths have a larger number of wave crests passing a particular point each second. Shorter-wavelength photons therefore pack more wiggles into a given interval of time, so will have higher frequencies—more wiggles per second. Each photon's frequency provides a direct measure of its energy: the higher the photon frequency, the more energy that photon carries.

As the cosmos cooled, photons lost energy to the expanding universe. The photons born in the gamma-ray and X-ray parts of the spectrum morphed into ultraviolet, visible light, and infrared photons. As their wavelengths grew larger, they became cooler and less energetic, but they never stopped being photons. Today, 13.7 billion years after the beginning, the photons of the cosmic background have shifted down the spectrum to become microwaves.

That's why astrophysicists call it the "cosmic microwave background," though a more enduring name is the "cosmic background radiation," or CBR. One hundred billion years from now, when the universe has expanded and cooled some more, future astrophysicists will describe the CBR as the "cosmic radio-wave background."

The temperature of the universe drops as the size of the universe grows. It's a physical thing. As different parts of the universe move apart, the wavelengths of the photons in the CBR must increase: the cosmos stretches these waves within the spandex fabric of space and time. Because every photon's energy varies in inverse proportion to its wavelength, all the free-traveling photons will lose half their original energy for every doubling in size of the cosmos.

All objects with temperatures above absolute zero will radiate photons throughout all parts of the spectrum. But that radiation always has a peak somewhere. The peak energy output of an ordinary household light bulb lies in the infrared part of the spectrum, which you can detect as warmth on your skin. Of course light bulbs also emit plenty of visible light, or we wouldn't buy them. So you can feel a lamp's radiation as well as see it.

The peak output of the cosmic background radiation occurs at a wavelength of about 1 millimeter, smack dab in the microwave part of the spectrum. The static that you hear on a walkie-talkie comes from an ambient bath of microwaves, a few percent of which are from the CBR. The rest of the "noise" comes from the Sun, cell phones, police radar guns, and so on. Besides peaking in the microwave region, the CBR also contains some radio waves (which allow it to contaminate Earth-based radio signals) and a vanishingly small number of photons with energies higher than those of microwaves.

The Ukrainian-born American physicist George Gamow and his colleagues predicted the existence of the CBR during the

1940s, consolidating their efforts in a 1948 paper that applied the
then-known laws of physics to the strange conditions of the early
universe. The foundation for their ideas came from the 1927
paper by Georges Edouard Lemaître, a Belgian astronomer and
Jesuit priest, now generally recognized as the "father" of big bang
cosmology. But two U.S. physicists, Ralph Alpher and Robert Herman, who had previously collaborated with Gamow, first estimated
what the temperature of the cosmic background ought to be.

In hindsight, Alpher, Gamow, and Herman had what today
seems a relatively simple argument, one which we have already
made: the fabric of space-time was smaller yesterday than it is
today, and since it was smaller, basic physics requires that it was
hotter. So the physicists turned back the clock to imagine the
epoch we have described, the time when the universe was so hot
that all its atomic nuclei were laid bare because photon collisions
knocked all electrons loose to roam freely through space. Under
those conditions, Alpher and Herman hypothesized, photons
could not have sped uninterrupted across the universe, as they do
today. The photons' current free ride requires that the cosmos
grew sufficiently cool for the electrons to settle into orbits around
the atomic nuclei. This formed complete atoms and allowed light
to travel without obstruction.

Although Gamow had the crucial insight that the early universe
must have been much hotter than our universe today, Alpher and
Herman were the first to calculate what its temperature would be
today: 5 degrees Kelvin. Yes, they got the number wrong—the
CBR actually has a temperature of 2.73 degrees Kelvin. But these
three guys nevertheless performed a successful extrapolation back
into the depths of long-vanished cosmic epochs—as great a feat as
any other in the history of science. To take some basic atomic
physics from a slab in the lab, and to deduce from it the largest-
scale phenomenon ever measured—the temperature history of
our universe—ranks as nothing short of mind-blowing. Assessing

this accomplishment, J. Richard Gott III, an astrophysicist at Princeton University, wrote in *Time Travel in Einstein's Universe*: "Predicting that the radiation existed and then getting its temperature correct to within a factor of 2 was a remarkable accomplishment—rather like predicting that a flying saucer 50 feet in width would land on the White House lawn and then watching one 27 feet in width actually show up."

When Gamow, Alpher, and Herman made their predictions, physicists were still undecided about the story of how the universe began. In 1948, the same year that Alpher and Herman's paper appeared, a rival "steady state" theory of the universe appeared in two papers published in England, one coauthored by the mathematician Hermann Bondi and the astrophysicist Thomas Gold, the other by the cosmologist Fred Hoyle. The steady state theory requires that the universe, though expanding, has always looked the same—a hypothesis with a deeply attractive simplicity. But because the universe is expanding, and because a steady state universe would not have been any hotter or denser yesterday than today, the Bondi-Gold-Hoyle scenario maintained that matter continuously pops into our universe at just the right rate to maintain a constant average density in the expanding cosmos. In contrast, the big bang theory (given its name in scorn by Fred Hoyle) requires that all matter come into existence at one instant, which some find more emotionally satisfying. Notice that the steady state theory takes the issue of the origin of the universe and throws it backward an infinite distance in time—highly convenient for those who would rather not examine this thorny problem.

The prediction of the cosmic background radiation amounted to a shot across the bow of the steady state theorists. The CBR's existence would clearly demonstrate that the universe was once far different—much smaller and hotter—from the way we find it today. The first direct observations of the CBR therefore put the

first nails in the coffin of the steady state theory (though Fred
Hoyle never fully accepted the CBR as disproving his elegant theory, going to his grave attempting to explain the radiation as arising from other causes). In 1964, the CBR was inadvertently and
serendipitously discovered by Arno Penzias and Robert Wilson at
the Bell Telephone Laboratories (Bell Labs, for short) in Murray
Hill, New Jersey. Little more than a decade later, Penzias and Wilson received the Nobel Prize for their good luck and hard work.

What led Penzias and Wilson to their Nobel Prize? During the
early 1960s, physicists all knew about microwaves, but almost no
one had created the capability of detecting weak signals in the
microwave portion of the spectrum. Back then, most wireless
communication (e.g., receivers, detectors, and transmitters) rode
on radio waves, which have longer wavelengths than microwaves.
For these, scientists needed a shorter-wavelength detector and a
sensitive antenna to capture them. Bell Labs had one, a king-size,
horn-shaped antenna that could focus and detect microwaves as
well as any apparatus on Earth.

If you're going to send or receive a signal of any kind, you don't
want other signals to contaminate it. Penzias and Wilson were
trying to open up a new channel of communication for Bell
Labs—so they wanted to pin down how much contaminating
"background" interference these signals would experience—from
the Sun, from the center of the galaxy, from terrestrial sources,
from whatever. They therefore embarked on a standard, important, and entirely innocent measurement, aimed at establishing
how easily they could detect microwave signals. Though Penzias
and Wilson had some astronomy background, they were not cosmologists but technophysicists studying microwaves, unaware of
the predictions made by Gamow, Alpher, and Herman. What they
were decidedly *not* looking for was the cosmic microwave background.

So they ran their experiment, and corrected their data for all known sources of interference. But they found background noise in the signal that didn't go away, and they couldn't figure out how to get rid of it. The noise seemed to come from every direction above the horizon, and it didn't change with time. Finally they looked inside their giant horn. Pigeons were nesting there, leaving a white dielectric substance (pigeon poop) everywhere nearby. Things must have been getting desperate for Penzias and Wilson: could the droppings, they wondered, be responsible for the background noise? They cleaned it up, and sure enough, the noise dropped a bit. But it still wouldn't go away. The paper they published in 1965 in *The Astrophysical Journal* refers to the persistent puzzle of an inexplicable "excess antenna temperature," rather than the astronomical discovery of the century.

While Penzias and Wilson were scrubbing bird droppings from their antenna, a team of physicists at Princeton University led by Robert H. Dicke was building a detector specifically designed to find the CBR that Gamow, Alpher, and Herman had predicted. The professors, however, lacked the resources of Bell Labs, so their work proceeded more slowly. The moment that Dicke and his colleagues heard about Penzias and Wilson's results, they knew that they'd been scooped. The Princeton team knew exactly what the "excess antenna temperature" was. Everything fit the theory: the temperature, the fact that the signal came from all directions in equal amounts, and that it wasn't linked in time with Earth's rotation or Earth's position in orbit around the Sun.

But why should anybody accept the interpretation? For good reason. Photons take time to reach us from distant parts of the cosmos, so we inevitably look back in time whenever we look outward into space. This means that if the intelligent inhabitants of a galaxy far, far away measured the temperature of the cosmic background radiation for themselves, long before we managed to so do, they should have found its temperature to be greater than 2.73 degrees Kelvin, because they would have inhabited the universe when it was younger, smaller, and hotter than it is today.

Can such an audacious assertion be tested? Yup. Turns out that the compound of carbon and nitrogen called cyanogen—best known to convicted murderers as the active ingredient of the gas administered by their executioners—will become excited by exposure to microwaves. If the microwaves are warmer than the ones in our CBR, they will excite the molecule a little more effectively than our microwaves do. The cyanogen compounds thus act as a cosmic thermometer. When we observe them in distant, and thus younger, galaxies, they should find themselves bathed in a warmer cosmic background than the cyanogen in our Milky Way galaxy. In other words, those galaxies ought to live more excited lives than we do. And they do. The spectrum of cyanogen in distant galaxies shows the microwaves to have just the temperature we expect at these earlier cosmic times.

You can't make this stuff up.

The CBR does far more for astrophysicists than to provide direct evidence for a hot early universe, and thus for the big bang model. It turns out that the details of the photons that comprise the CBR reach us laden with information about the cosmos both before and after the universe became transparent. We have noted that until that time, about 380,000 years after the big bang, the universe was opaque, so you couldn't have witnessed matter making shapes even if you'd been sitting front-row center. You couldn't have seen where galaxy clusters were starting to form. Before anybody, anywhere, could see anything worth seeing, photons had to acquire the ability to travel, unimpeded, across the universe. When the time was right, each photon began its cross-cosmos journey at the point where it smacked into the last electron that would ever stand in its way. As more and more photons escaped without being deflected by electrons (thanks to electrons joining nuclei to form atoms) they created an expanding shell of photons that astrophysicists call "the surface of last scatter." That shell, which formed during a period of about a hundred thousand years, marks the epoch when almost all the atoms in the cosmos were born.

By then, matter in large regions of the universe had already begun to coalesce. Where matter accumulates, gravity grows stronger, enabling more and more matter to gather. Those matter-rich regions seeded the formation of galaxy superclusters, while other regions remained relatively empty. The photons that last scattered off electrons within the coalescing regions developed a different, slightly cooler spectrum as they climbed out of the strengthening gravity field, which robbed them of a bit of energy.

The CBR indeed shows spots that are slightly hotter or slightly cooler than average, typically by about one hundred-thousandth of a degree. These hot and cool spots mark the earliest structures in the cosmos, the first clumping together of matter. We know what matter looks like today because we see galaxies, galaxy clusters, and galaxy superclusters. To figure out how those systems arose, we probe the cosmic background radiation, a remarkable relic from the remote past, still filling the entire universe. Studying the patterns in the CBR amounts to a kind of cosmic phrenology: we can read the bumps on the "skull" of the youthful universe and from them deduce behavior not only for an infant but also for a grown-up.

By adding other observations of the local and the distant universe, astronomers can determine all kinds of fundamental cosmic properties from the CBR. Compare the distribution of sizes and temperatures of the slightly warmer and cooler areas, for instance, and we can infer the strength of gravity in the early universe, and thus how quickly matter accumulated. From that we can then deduce how much ordinary matter, dark matter, and dark energy the universe comprises (the percentages are 4, 23, and 73, respectively). From there, it's easy to tell whether or not the universe will expand forever, and whether or not the expansion will slow down or speed up as time passes.

Ordinary matter is what everyone is made of. It exerts gravity and can absorb, emit, and otherwise interact with light. Dark matter, as we'll see in Chapter 4, is a substance of unknown nature that produces gravity but does not interact with light in any known way. And dark energy, as we'll see in Chapter 5, induces an acceleration of the cosmic expansion, forcing the universe to expand more rapidly than it otherwise would. The phrenology exam now says that cosmologists understand how the early universe behaved, but that most of the universe, then and now, consists of stuff they're clueless about.

Profound areas of ignorance notwithstanding, today, as never before, cosmology has an anchor. The CBR carries the imprint of a portal through which all of us once passed.

The discovery of the cosmic microwave background added new precision to cosmology by verifying the conclusion, originally derived from observations of distant galaxies, that the universe has been expanding for billions of years. It was the accurate and detailed map of the CBR—a map first made for small patches of the sky using balloon-borne instruments and a telescope at the South Pole, and then for the entire sky by a satellite called the Wilkinson Microwave Anisotropy Probe (WMAP)—that secured cosmology's place at the table of experimental science. We shall hear much more from WMAP, whose first results appeared in 2003, before our cosmological tale is done.

Cosmologists have plenty of ego: how else could they have the audacity to deduce what brought the universe into being? But the new era of observational cosmology may call for a more modest, less freewheeling stance among its practitioners. Each new observation, each morsel of data, can be good or bad for your theories.

On the one hand, the observations provide a basic foundation for cosmology, a foundation that so many other sciences can take for granted because they achieve rich streams of laboratory observations. On the other hand, new data will almost certainly dispatch some of the tall tales that theorists dreamed up when they lacked the observations that would give them thumbs up or down.

No science achieves maturity without precision data. Cosmology has now become precision science.

CHAPTER 4

Let There Be Dark

Gravity, the most familiar of nature's forces, offers us simultaneously the best and the least understood phenomena in nature. It took the mind of Isaac Newton, the millennium's most brilliant and influential, to realize that gravity's mysterious "action at a distance" arises from the natural effects of every bit of matter, and that the attractive forces between any two objects can be described by a simple algebraic equation. It took the mind of Albert Einstein, the twentieth century's most brilliant and influential, to show that we can more accurately describe gravity's action-at-a-distance as a warp in the fabric of space-time, produced by any combination of matter and energy. Einstein demonstrated that Newton's theory requires some modification to describe gravity accurately—in predicting, for example, the amount by which light rays will bend when they pass by a massive object. Although Einstein's equations are fancier than Newton's, they nicely accommodate the matter that we have come to know and love. Matter that we can see, touch, feel, and occasionally taste.

Don't know who's next in the genius sequence, but we've now been waiting well over half a century for somebody to tell us why the bulk of all the gravitational forces that we've measured in the universe arises from substances that we have neither seen, nor touched, nor felt, nor tasted. Or maybe the excess gravity doesn't come from matter at all, but emanates from some other conceptual thing. In any case, we are without a clue. We find ourselves no closer to an answer today than we were when this "missing mass" problem was first identified in 1933 by astronomers who measured the velocities of galaxies whose gravity affects their close neighbors, and more fully analyzed in 1937 by the colorful Bulgarian-Swiss-American astrophysicist Fritz Zwicky, who taught at the California Institute of Technology for more than forty years, combining his far-ranging insights into the cosmos with a colorful means of expression and an impressive ability to antagonize his colleagues.

Zwicky studied the movement of galaxies within a titanic cluster of galaxies, located far beyond the local stars of the Milky Way that trace out the constellation Coma Berenices (the "hair of Berenice," an Egyptian queen in antiquity). The Coma cluster, as it is called by those in the know, is an isolated and richly populated ensemble of galaxies about 300 million light-years from Earth. Its many thousands of galaxies orbit the cluster's center, moving in all directions like bees circling their hive. Using the motions of a few dozen galaxies as tracers of the gravity field that binds the entire cluster, Zwicky discovered their average velocity to be shockingly high. Since larger gravitational forces induce higher velocities in the objects that they attract, Zwicky deduced an enormous mass for the Coma cluster. When we sum up all of its galaxies' estimated masses, Coma ranks among the largest and most massive galaxy clusters in the universe. Even so, the cluster does not contain enough visible matter to account for the observed speeds of its member galaxies. Matter seems to be missing.

If you apply Newton's law of gravity and assume that the cluster does not exist in an odd state of expansion or collapse, you can calculate what the characteristic average speed of its galaxies ought to be. All you need is the size of the cluster and an estimate of its total mass: The mass, acting over distances characterized by the cluster's size, determines how rapidly the galaxies must move to avoid falling into the cluster's center or escaping from the cluster entirely.

In a similar calculation, as Newton showed, you can derive the speed at which each of the planets at its particular distance from the Sun must move in its orbit. Far from being magic, these speeds satisfy the gravitational circumstance in which each planet finds itself. If the Sun suddenly acquired more mass, Earth and everything else in the solar system would need larger velocities to stay in their current orbits. With too much speed, however, the Sun's gravity will be insufficient to maintain everybody's orbit. If Earth's orbital speed were more than the square root of 2 times its current speed, our planet would achieve "escape velocity" and, you guessed it, escape the solar system. We can apply the same reasoning to much larger objects, such as our own Milky Way galaxy, in which stars move in orbits that respond to the gravity from all the other stars, or in clusters of galaxies, where each of the galaxies likewise feels the gravity from all the other galaxies. As Einstein once wrote (more ringingly in German than in this English translation by one of us [DG]) to honor Isaac Newton:

Look unto the stars to teach us
How the master's thoughts can reach us
Each one follows Newton's math
Silently along its path.

When we examine the Coma cluster, as Zwicky did during the 1930s, we find that its member galaxies all move more rapidly than the escape velocity for the cluster, but only if we establish that velocity from the sum of all the galaxy masses taken one by one, which we estimate from the galaxies' brightnesses. The cluster should therefore swiftly fly apart, leaving barely a trace of its beehive existence after just a few hundred million years, perhaps a billion, had passed. But the cluster is more than 10 billion years old, nearly as old as the universe itself. And so was born what remains the longest-standing mystery in astronomy.

Through the decades that followed Zwicky's work, other galaxy clusters revealed the same problem. So Coma could not be blamed for being odd. Then whom should we blame? Newton? No, his theories had been examined for 250 years and passed all tests. Einstein? No. The formidable gravity of galaxy clusters does not rise high enough to require the full hammer of Einstein's general theory of relativity, just two decades old when Zwicky did his research. Perhaps the "missing mass" needed to bind the Coma cluster's galaxies does exist, but in some unknown, invisible form. For a time, astronomers renamed the missing-mass problem the "missing-light problem," since the mass had been strongly inferred from the excess of gravity. Today, with better determinations of the masses of galaxy clusters, astronomers use the moniker "dark matter," although "dark gravity" would be more precise.

The dark matter problem reared its invisible head a second time. In 1976, Vera Rubin, an astrophysicist at the Carnegie Institution of Washington, discovered a similar "missing-mass" anomaly within spiral galaxies themselves. Studying the speeds at which stars orbit their galaxy centers, Rubin first found what she expected: within the visible disk of each galaxy, the stars farther from the center move at greater speeds than stars close in. The farther stars have more matter (stars and gas) between themselves and the galaxy center, requiring higher speeds to sustain their orbits. Beyond the galaxy's luminous disk, however, we can still find some isolated gas clouds and a few bright stars. Using these objects as tracers of the gravity field "outside" the galaxy, where visible matter no longer adds to the total, Rubin discovered that their orbital speeds, which should have fallen with increasing distance out there in Nowheresville, in fact remained high.

These largely empty volumes of space—the rural regions of each galaxy—contain too little visible matter to explain the orbital speeds of the tracers. Rubin correctly reasoned that some form of dark matter must lie in these far-out regions, well beyond the visible edge of each spiral galaxy. Indeed, the dark matter forms a kind of halo around the entire galaxy.

This halo problem exists under our noses, right in our own Milky Way galaxy. From galaxy to galaxy and from cluster to cluster, the discrepancy between the mass in visible objects and the total mass of systems ranges from a factor of just 2 or 3 up to factors of many hundreds. Across the universe, the factor averages to about 6. That is, cosmic dark matter enjoys about six times the mass of all the visible matter.

Over the past twenty-five years, further research has revealed that most of the dark matter cannot consist of nonluminous ordinary matter. This conclusion rests on two lines of reasoning. First, we can eliminate with near certainty all plausible familiar candidates, like the suspects in a police lineup: Could the dark matter reside in black holes? No, we think that we would have detected this many black holes from their gravitational effects on nearby stars. Could it be dark clouds? No, they would absorb or otherwise interact with light from stars behind them, which real dark matter doesn't do. Could it be interstellar (or intergalactic) planets, asteroids, and comets, all of which produce no light of their own? It's hard to believe that the universe would manufacture six times as much mass in planets as in stars. That would mean six thousand Jupiters for every star in the galaxy, or worse yet, 2 million

Earths. In our own solar system, for example, everything that is not the Sun adds to a paltry 0.2 percent of the Sun's mass.

Thus, as best we can figure, the dark matter doesn't simply consist of matter that happens to be dark. Instead, it's something else altogether. Dark matter exerts gravity according to the same rules that ordinary matter follows, but it does little else that might allow us to detect it. Of course, we are hamstrung in this analysis by not knowing what the dark matter is. The difficulties of detecting dark matter, intimately connected with our difficulties in perceiving what it might be, raise the question: If all matter has mass, and all mass has gravity, does all gravity have matter? We don't know. The name "dark matter" presupposes the existence of a kind of matter that has gravity and that we don't yet understand. But maybe it's the gravity that we don't understand.

To study dark matter beyond deducing its existence, astrophysicists now seek to learn where the stuff collects in space. If dark matter existed only at the outer edges of galaxy clusters, for example, then the galaxies' velocities would show no evidence of a dark matter problem, because the galaxies' speeds and trajectories respond only to sources of gravity interior to their orbits. If the dark matter occupied only the clusters' centers, then the run of galaxy speeds as measured from the center of the cluster out to its edge would respond to ordinary matter alone. But the speeds of galaxies in clusters reveal that the dark matter permeates the entire volume occupied by the orbiting galaxies. In fact, the locations of ordinary matter and dark matter loosely coincide with each other. Several years ago, a team led by the American astrophysicist J. Anthony Tyson, then at Bell Labs and now at UC Davis (he's called "Cousin Tony" by one of us, though we have no family relationship) produced the first detailed map of the distribution of dark matter's gravity in and around a titanic cluster of galaxies. Wherever we see big galaxies, we also find a higher concentration of dark matter within the cluster. The converse is also true: regions with no visible galaxies have a dearth of dark matter.

The discrepancy between dark and ordinary matter varies significantly from one astrophysical environment to another, but it becomes most pronounced for large entities such as galaxies and galaxy clusters. For the smallest objects, such as moons and planets, no discrepancy exists. Earth's surface gravity, for example, can be explained entirely by what's under our feet. So if you are overweight on Earth, don't blame dark matter. Dark matter also has no bearing on the Moon's orbit around Earth, nor on the movements of the planets around the Sun. But we do need it to explain the motions of stars around the center of the galaxy.

Does a different kind of gravitational physics operate on the galactic scale? Probably not. More likely, dark matter consists of matter whose nature we have yet to divine, and which clusters more diffusely than ordinary matter does. Otherwise, we would find that one in every six pieces of dark matter has a chunk of ordinary matter clinging to it. So far as we can tell, that's not the way things are.

At the risk of inducing depression, astrophysicists sometimes argue that all the matter that we have come to know and love in the universe—the stuff of stars, planets, and life—are mere buoys afloat in a vast cosmic ocean of something that looks like nothing.

But what if this conclusion were entirely wrong? When nothing else seems to work, some scientists will understandably, and quite rightly, question the fundamental laws of physics that underlies the assumptions made by others who seek to understand the universe.

During the early 1980s, the Israeli physicist Mordehai Milgrom of the Weizmann Institute of Science in Rehovot, Israel, suggested a change in Newton's laws of gravity, a theory now known as MOND (MOdified Newtonian Dynamics). Accepting the fact that standard Newtonian dynamics operates successfully on size scales smaller than galaxies, Milgrom suggested that Newton needed some help in describing gravity's effects at distances the sizes of galaxies and galaxy clusters, within which individual stars and star clusters are so far apart that they exert relatively little gravitational force on each other. Milgrom added an extra term to Newton's equation, specifically tailored to come to life at astronomically large distances. Although he invented MOND as a computational tool, Milgrom didn't rule out the possibility that his theory could refer to a new phenomenon of nature.

MOND has had only limited success. The theory can account for the movement of isolated objects in the outer reaches of many spiral galaxies, but it raises more questions than it answers. MOND fails to predict reliably the dynamics of more complex configurations, such as the movement of galaxies in binary and multiple systems. Furthermore, the detailed map of the cosmic background radiation produced by the WMAP satellite in 2003 allowed cosmologists to isolate and measure the influence of dark matter in the early universe. Because these results appear to correspond to a consistent model of the cosmos based on conventional theories of gravity, MOND has lost many adherents.

During the first half million years after the big bang, a mere moment in the 14-billion-year sweep of cosmic history, matter in the universe had already begun to coalesce into the blobs that would become clusters and superclusters of galaxies. But the cosmos was expanding all along, and would double in size during its next half million years. So the universe responds to two competing effects: gravity wants to make stuff coagulate, but the expansion wants to dilute it. If you do the math, you rapidly deduce that the gravity from ordinary matter could not win this battle by itself. It needed the help of dark matter, without which we would be living—actually not living—in a universe with no structure: no clusters, no galaxies, no stars, no planets, no people. How much gravity from dark matter did it need? Six times as much as that provided by ordinary matter itself. This analysis leaves no room for MOND's little corrective terms in Newton's laws. The analysis doesn't tell us what dark matter is, only that dark matter's effects are real—and that, try as you may, you cannot credit ordinary matter for it.

Dark matter plays another crucial role in the universe. To appreciate all that the dark matter has done for us, go back in time to a couple of minutes after the big bang, when the universe was still so immensely hot and dense that hydrogen nuclei (protons) could fuse together. This crucible of the early cosmos forged hydrogen into helium, along with trace amounts of lithium, plus an even smaller amount of deuterium, which is a heavier version of the hydrogen nucleus, with a neutron added to the proton. This mixture of nuclei provides another cosmic fingerprint of the big bang, a relic that allows us to reconstruct what happened when the cosmos was a few minutes old. In creating this fingerprint, the prime mover was the strong nuclear force—the force that binds protons and neutrons within the nucleus—and not gravity, a force so weak that it gains significance only as particles assemble themselves by the trillions.

By the time the temperature dropped below a threshold value, nuclear fusion throughout the universe had made one helium nucleus for every ten hydrogen nuclei. The universe had also turned about one part in a thousand of its ordinary matter into lithium nuclei, and two parts in a hundred thousand into deuterium. If dark matter did not consist of some noninteracting substance but was instead made of dark ordinary matter—matter with normal fusion privileges—then because the dark matter packed six times as many particles into the tiny volumes of the early universe as ordinary matter did, its presence would have dramatically increased the fusion rate of hydrogen. The result would have been a noticeable overproduction of helium, in comparison with the observed amount, and the birth of a universe notably different to the one that we inhabit.

Helium is one tough nucleus, relatively easy to make but extremely difficult to fuse into other nuclei. Because stars have continued to make helium from hydrogen in their cores, while destroying relatively little helium through more advanced nuclear fusion, we may expect that the places where we find the lowest amounts of helium in the universe should have no less helium than what the universe produced during its first few minutes. Sure enough, galaxies whose stars have only minimally processed their ingredients show that one in ten of their atoms consists of helium, just as you would expect from the big bang birthday suit of the cosmos, so long as the dark matter then present did not participate in the nuclear fusion that created nuclei.

So, dark matter is our friend. But astrophysicists understandably grow uncomfortable whenever they must base their calculations on concepts they don't understand, even though this wouldn't be the first time they've done so. Astrophysicists measured the energy output of the Sun, for instance, long before anybody knew that thermonuclear fusion was responsible. Back in the nineteenth century, before the introduction of quantum mechanics and the discovery of other deep insights into the behavior of matter on its smallest scales, fusion didn't even exist as a concept.

Unrelenting skeptics might compare the dark matter of today with the hypothetical, now defunct "ether," proposed centuries ago as the weightless, transparent medium through which light moved. For many years, until a famous 1887 experiment in Cleveland performed by Albert Michelson and Edward Morley, physicists assumed that the ether must exist, even though not a shred of evidence supported this presumption. Known to be a wave, light was thought to require a medium through which to move, much as sound waves move through air. Light turns out to be quite happy, however, traveling through the vacuum of space, devoid of any supporting medium. Unlike sound waves, however, which do consist of vibrations of the air, light waves propagate themselves.

But dark matter ignorance differs fundamentally from ether ignorance. While ether amounted to a placeholder for our incomplete understanding, the existence of dark matter derives not from mere presumption but from the observed effects of its gravity on visible matter. We're not inventing dark matter out of thin space; instead, we deduce its existence from observational facts. Dark matter is just as real as the hundred-plus planets discovered in orbit around stars other than the Sun—almost all of them found solely by their gravitational influence on their host stars. The worst that can happen is that physicists (or others of deep insight) might discover that the dark matter does not consist of matter at all, but of something else, yet they cannot argue it away. Could dark matter be the manifestation of forces from another dimension? Or of a parallel universe intersecting ours? Even so, none of this would change the successful invocation of dark matter's gravity in the equations that we use to understand the formation and evolution of the universe.

Other unrelenting skeptics might declare that "seeing is believing." A seeing-is-believing approach to life works well in many endeavors, including mechanical engineering, fishing, and perhaps dating. It's also good, apparently, for residents of Missouri. But it doesn't make for good science. Science is not just about seeing. Science is about measuring—preferably with something that's *not* your own eyes, which are inextricably conjoined with the baggage of your brain: preconceived ideas, post-conceived notions, imagination unchecked by reference to other data, and bias.

Having resisted attempts to detect it directly on Earth for three quarters of a century, dark matter has become a type of Rorschach test of the investigator. Some particle physicists say the dark matter must consist of some ghostly class of undiscovered particles that interact with matter via gravity, but otherwise interact with matter or light only weakly, or not at all. This sounds off-the-wall, but the suggestion has precedent. Neutrinos, for instance, are well known to exist, though they interact extremely weakly with ordinary light and matter. Neutrinos from the Sun— two neutrinos for every helium nucleus made in the solar core— travel through the vacuum of space at nearly the speed of light, but then pass through Earth as though it did not exist. The tally: night and day, 100 billion neutrinos from the Sun enter, then exit each square inch of your body every second.

But neutrinos can be stopped. Every rare once in a while they interact with matter via nature's weak nuclear force. And if you can stop a particle, you can detect it. Compare neutrinos' elusive behavior with that of the Invisible Man (in his invisible phase)— as good a candidate for the dark matter as anything else. He could walk through walls and doors as though they were not there. Although equipped with these talents, why didn't he didn't just drop through the floor into the basement?

If we can build sufficiently sensitive detectors, the particle physicist's dark-matter particles may reveal themselves through familiar interactions. Or they may reveal their presence through forces other than the strong nuclear force, the weak nuclear force, and electromagnetism. These three forces (plus gravity) mediate all interactions between and among all known particles. So the choices are clear. Either dark matter particles must wait for us to discover and to control a new force or class of forces through which the particles interact, or else dark matter particles interact via normal forces, but with staggering weakness.

MOND theorists see no exotic particles in their Rorschach tests.

They think gravity, not particles, is what needs fixing. And so they brought forth modified Newtonian dynamics—a bold attempt that seems to have failed, but doubtless the precursor of other efforts to change our view of gravity rather than our census of subatomic particles.

Other physicists pursue what they call TOEs or "theories of everything." In a spin-off of one version, our own universe indeed lies near a parallel universe, with which we interact only through gravity. You'll never run into any matter from that parallel universe, but you might feel its tug, crossing into the spatial dimensions of our own universe. Imagine a phantom universe right next to ours, revealed to us only through its gravity. Sounds exotic and unbelievable, but probably not any more so than the first suggestions that Earth orbits the Sun, or that our galaxy is not alone in the universe.

So, dark matter's effects are real. We just don't know what the dark matter is. It seems not to interact through the strong force, so it cannot make nuclei. It hasn't been found to interact through the weak nuclear force, something even elusive neutrinos do. It doesn't seem to interact with the electromagnetic force, so it doesn't make molecules, or absorb or emit or reflect or scatter light. It does exert gravity, however, to which ordinary matter responds. That's it. After all these years of investigation, astrophysicists haven't discovered it doing anything else.

Detailed maps of the cosmic background radiation have demonstrated that dark matter must have existed during the first 380,000 years of the universe. We also need dark matter today in our own galaxy and in galaxy clusters to explain the motions of objects they contain. But as far as we know, the march of astrophysics has not yet been derailed or stymied by our ignorance. We simply carry dark matter along as a strange friend, and invoke it where and when the universe requires it of us.

In what we hope is the not-so-distant future, the fun will continue as we learn to exploit dark matter—once we figure out what the stuff is made of. Imagine invisible toys, cars that pass through one another, or super stealth airplanes. The history of obscure and obtuse discoveries in science is rich with examples of clever people who came later and who figured out how to exploit such knowledge for their own gain or for the benefit of life on Earth.

CHAPTER 5

Let There Be More Dark

The cosmos, we now know, has both a light and a dark side. The light side embraces all familiar heavenly objects—the stars, which group by the billions into galaxies, as well as the planets and smaller cosmic debris that may not produce visible light but do emit other forms of electromagnetic radiation, such as infrared or radio waves.

We have discovered that the dark side of the universe embraces the puzzling dark matter, detected only by its gravitational influence on visible matter but otherwise of completely unknown form and composition. A modest amount of this dark matter may be ordinary matter that remains invisible because it produces no detectable radiation. But, as detailed in the previous chapter, the great bulk of the dark matter must consist of non-ordinary matter, whose nature continues to elude us—except for its gravitational force on matter we can see.

Beyond all issues concerning dark matter, the dark side of the universe has another, entirely different aspect. One that resides not in matter of any kind, but in space itself. We owe this concept, along with the amazing results that it implies, to the father of modern cosmology, none other than Albert Einstein himself.

Ninety years ago, while the newly perfected machine guns of World War I slaughtered soldiers by the thousands a few hundred miles to the west, Albert Einstein sat in his office in Berlin, pondering the universe. As the war began, Einstein and a colleague had circulated an antiwar petition among his peers, gathering two other signatures in addition to their own. This act set him apart from his fellow scientists, most of whom had signed an appeal to aid Germany's war effort, and ruined his colleague's career. But Einstein's engaging personality and scientific fame allowed him to keep the esteem of his peers. He continued his efforts to find equations that could accurately describe the cosmos.

Before the war ended, Einstein achieved success—arguably his greatest of all. In November 1915, he produced his general theory of relativity, which describes how space and matter interact: Matter tells space how to bend, and space tells matter how to move. To replace Isaac Newton's mysterious "action at a distance," Einstein viewed gravity as a local warp in the fabric of space. The Sun, for example, creates a sort of dimple, bending space most noticeably at distances closest to it. The planets tend to roll into this dimple, but their inertia keeps them from falling all the way in. Instead, they move in orbits around the Sun that keep them at a nearly constant distance from the dimple in space. Within a few weeks after Einstein published his theory, the physicist Karl Schwarzschild, diverting himself from the horrors of life in the German army (which gave him a fatal disease soon afterward), used Einstein's concept to demonstrate that an object with sufficiently strong gravity will create a "singularity" in space. At such a singularity, space bends completely around the object and prevents anything, including light, from leaving its immediate vicinity. We now call these objects black holes.

Einstein's theory of general relativity led him to the key equation he had been seeking, one that links the contents of space to its overall behavior. Studying this equation in the privacy of his office, creating models of the cosmos in his mind, Einstein almost discovered the expanding universe, a dozen years before Edwin Hubble's observations revealed it.

Einstein's basic equation predicts that in a universe in which matter has a roughly even distribution, space cannot be "static." The cosmos cannot just "sit there," as our intuition insists that it should, and as all astronomical observations until that time implied. Instead, the totality of space must always be either expanding or contracting: space must behave something like the surface of an inflating or deflating balloon, but never like the surface of a balloon with constant size.

This worried Einstein. For once, this bold theorist, who mistrusted authority and had never hesitated to oppose conventional physics ideas, felt that he had gone too far. No astronomical observations suggested an expanding universe, because astronomers had only documented the motions of nearby stars and had not yet determined the faraway distances to what we now call galaxies. Rather than announcing to the world that the universe must either be expanding or contracting, Einstein returned to his equation, seeking a way to immobilize the cosmos.

He soon found one. Einstein's basic equation allowed for a term with a constant but unknown value that represents the amount of energy contained in every cubic centimeter of empty space. Because nothing suggested that this constant term should have one value or another, in his first pass Einstein had set it equal to zero. Now Einstein published a scientific article to demonstrate that if this constant term, which cosmologists later named the "cosmological constant," had a particular value, then space could be static. Then theory would no longer conflict with observations of the universe, and Einstein could regard his equation as valid.

Einstein's solution encountered grave difficulties. In 1922, a Russian mathematician named Alexander Friedmann proved that Einstein's static universe must be unstable, like a pencil balanced on its point. The slightest ripple or disturbance would cause space either to expand or to contract. Einstein first proclaimed Friedmann mistaken, but then, in a generous act typical of the man, published an article retracting that claim and pronouncing Friedmann correct after all. As the 1920s ended, Einstein was delighted to learn of Hubble's discovery that the universe is expanding. According to George Gamow's recollections, Einstein pronounced the cosmological constant his "greatest blunder." Except for a few cosmologists who continued to invoke a non-zero cosmological constant (with a value different from the one that Einstein had used) to explain certain puzzling observations, most of which later proved to be incorrect, scientists the world over sighed with relief that space had proven to have no need of this constant.

Or so they thought. The great cosmological story at the end of the twentieth century, the surprise that stood the world of cosmology on one ear and sang a different tune into the other, resides in the stunning discovery, first announced in 1998, that the universe does have a non-zero cosmological constant. Empty space does indeed contain energy, named "dark energy," and possesses highly unusual characteristics that determine the future of the entire universe.

To understand, and possibly even to believe, these dramatic assertions, we must follow the crucial themes in cosmologists' thinking during the seventy years following Hubble's discovery of the expanding universe. Einstein's fundamental equation allows for the possibility that space can have curvature, described mathematically as positive, zero, or negative. Zero curvature describes "flat space," the kind that our minds insist on as the only possibility, which extends to infinity in all directions, like the surface of an infinite chalkboard. In contrast, a positively curved space corresponds in analogy to the surface of a sphere, a two-dimensional space whose curvature we can see by using the third dimension. Notice that the center of the sphere, the point that appears to remain stationary as its two-dimensional surface expands or contracts, resides in this third dimension and appears nowhere on the surface that represents all of space.

Just as all positively curved surfaces include only a finite amount of area, all positively curved spaces contain only a finite amount of volume. A positively curved cosmos has the property that if you journey outward from Earth for a sufficiently long time, you will eventually return to your point of origin, like Magellan circumnavigating our globe. Unlike positively curved spherical surfaces, negatively curved spaces extend to infinity, even though they are not flat. A negatively curved two-dimensional surface resembles the surface of an infinitely large saddle: it curves "upward" in one direction (front to back) and "downward" in another (side to side).

If the cosmological constant equals zero, we can describe the overall properties of the universe with just two numbers. One of these, called the Hubble constant, measures the rate at which the universe is expanding now. The other measures the curvature of space. During the second half of the twentieth century, almost all cosmologists believed that the cosmological constant was zero, and saw measuring the cosmic expansion rate and the curvature of space as their primary research agenda.

Both of these numbers can be found from accurate measurements of the speeds at which objects located at different distances are receding from us. The overall trend between distance and velocity—the rate at which the recession velocities of galaxies increase with increasing distance—yields the Hubble constant, whereas small deviations from this general trend, which appear only when we observe objects at the greatest distances from us, will reveal the curvature of space. Whenever astronomers observe objects many billion light-years from the Milky Way, they look so far back in time that they see the cosmos not as it is now but as it was when significantly less time had elapsed since the big bang. Observations of galaxies located 5 billion or more light-years from the Milky Way allow cosmologists to reconstruct a significant part of the history of the expanding universe. In particular, they can see how the rate of expansion has changed with time— the key to determining the curvature of space. This approach works, at least in principle, because the amount of space's curvature induces subtle differences in the rate at which the universal expansion had changed through past billions of years.

In practice, astrophysicists remained unable to fulfill this program, because they could not make sufficiently reliable estimates of the distances to galaxy clusters many billion light-years from Earth. They had another arrow in their quiver, however. If they could measure the average density of all the matter in the universe—that is, the average number of grams of material per cubic centimeter of space—they could compare this number with the "critical density," a value predicted by Einstein's equations that describe the expanding universe. The critical density specifies the exact density required for a universe with zero curvature of space. If the actual density lies above this value, the universe has positive curvature. In that case, assuming that the cosmological constant equals zero, the cosmos will eventually cease expanding and start contracting. If, however, the actual density exactly equals the critical density, or falls below it, then the universe will expand forever. Exact equality of the actual and critical values of the density occurs in a cosmos with zero curvature, whereas in a negatively curved universe, the actual density is less than the critical density.

By the mid-1990s, cosmologists knew that even after including all the dark matter they had detected (from its gravitational influence on visible matter), the total density of matter in the universe only came to about one quarter of the critical density. This result hardly seems astounding, although it does imply that the cosmos will never cease expanding, and that the space in which all of us live must be negatively curved. It hurt theoretically oriented cosmologists, however, because they had come to believe that space must have zero curvature.

This belief rested on the "inflationary model" of the universe, named (unsurprisingly) at a time of a steeply rising consumer price index. In 1979, Alan Guth, a physicist working at the Stanford Linear Accelerator Center in California, hypothesized that during its earliest moments, the cosmos expanded at an incredibly rapid rate—so rapidly that different bits of matter accelerated away from one another, reaching speeds far greater than the speed of light. But doesn't Einstein's theory of special relativity make the speed of light a universal speed limit for all motion? Not exactly. Einstein's limit applies only to objects moving within space and not the expansion of space itself. During the "inflationary epoch," which lasted only from about 10^{-37} second to 10^{-34} second after the big bang, the cosmos expanded by a factor of about 10^{50}.

What produced this enormous cosmic expansion? Guth speculated that all of space must have undergone a "phase transition," something analogous to what happens when liquid water quickly freezes into ice. After some crucial tweaking by his colleagues in the Soviet Union, the United Kingdom, and the United States, Guth's idea became so attractive that it has dominated theoretical models of the extremely early universe for two decades.

And what makes inflation such an attractive theory? The inflationary era explains why the universe, in its overall properties, looks the same in all directions: everything that we can see (and a good deal more than that) inflated from a single tiny region of space, converting its local properties into universal ones. Other advantages, which need not detain us here, accrue to the theory, at least among those who create model universes in their minds. One additional feature deserves emphasis, however. The inflationary model makes a straightforward, testable prediction: space in the universe should be flat, neither positively nor negatively curved, but just as flat as our intuition imagines it.

According to this theory, the flatness of space arises from the enormous expansion that occurred during the inflationary epoch. Picture yourself, in analogy, on the surface of a balloon, and let the balloon expand by a factor so large that you lose track of the zeros. After this expansion, the part of the balloon's surface that you can see will be flat as a pancake. So too should be all the space that we can ever hope to measure—if the inflationary model actually describes the real universe.

But the total density of matter amounts to only about one quarter of the amount required to make space flat. During the 1980s and 1990s, many theoretically minded cosmologists believed that because the inflationary model must be valid, new data would eventually close the cosmic "mass gap," the difference between the total density of matter, which pointed toward a negatively curved universe, and the critical density, seemingly required to achieve a cosmos with flat space. Their beliefs carried them buoyantly onward, even as observationally oriented cosmologists mocked their overreliance on theoretical analysis. And then the mocking stopped.

In 1998, two rival teams of astronomers announced new observations implying the existence of a non-zero cosmological constant—not (of course) the very number that Einstein had proposed in order to keep the universe static, but another, quite different value, one that implies that the universe will expand forever at an ever-increasing rate.

If theorists had proposed this for yet another model universe, the world would have little noted nor long remembered their effort. Here, however, reputable experts in observing the real universe, had mistrusted one another, checked on their rivals' suspicious activities, and discovered that they agreed on the data and their interpretation. The observational results not only implied a cosmological constant different from zero but also assigned to that constant a value that makes space flat.

What's that you say? The cosmological constant flattens space? Aren't you suggesting, like the Red Queen in *Alice in Wonderland*, that we each believe six impossible things before breakfast? More mature reflection may, however, convince you that if apparently empty space does contain energy (!), that energy must contribute mass to the cosmos, just as Einstein's famous equation, $E = mc^2$, implies. If you've got some E, you can conceive it as a corresponding amount of m, equal to E divided by c^2. Then the total density must equal the total of the density contributed by matter, plus the density contributed by energy.

The new total density is what we must compare with the critical density. If the two are equal, space must be flat. This would satisfy the inflationary model's prediction of flat space, for it does not care whether the total density in space arises from the density of matter, or the matter equivalent provided by the energy in empty space, or a combination of the two.

The crucial evidence suggesting a non-zero cosmological constant, and thus the existence of dark energy, came from astronomers' observations of a particular type of exploding star or supernova, stars that die spectacular deaths in titanic explosions. These supernovae, called either Type Ia or SN Ia's, differ from other types, which occur when the cores of massive stars collapse after exhausting all possibilities of producing more energy by nuclear fusion. In contrast, SN Ia's owe their origin to white dwarf stars that belong to binary star systems. Two stars that happen to be born close to one another will spend their lives performing simultaneous orbits around their common center of mass. If one of the two stars has more mass than the other, it will pass more rapidly through its prime of life, and in most cases will then lose its outer layers of gas, revealing its core to the cosmos as a shrunken, degenerate "white dwarf," an object no larger than Earth but containing as much mass as the Sun. Physicists call the matter in white dwarfs "degenerate" because it has such a high density— more than a hundred thousand times the density of iron or gold—that the effects of quantum mechanics act on matter in bulk form, preventing it from collapsing under its enormous self-gravitational forces.

A white dwarf in mutual orbit with an aging companion star attracts gaseous material that escapes from the star. This matter, still relatively rich in hydrogen, accumulates on the white dwarf, growing steadily denser and hotter. Finally, when the temperature rises to 10 million degrees, the entire star ignites in nuclear fusion. The resulting explosion—similar in concept to a hydrogen bomb but trillions of times more violent—blows the white dwarf completely apart and produces a Type Ia supernova.

SN Ia's have proven particularly useful to astronomers by possessing two separate qualities. First, they produce the most luminous supernova explosions in the cosmos, visible across billions of light-years. Second, nature sets a limit to the maximum mass that any white dwarf can have, equal to about 1.4 times the Sun's mass. Matter can accumulate on a white dwarf's surface only until the white dwarf's mass reaches this limiting value. At that point, nuclear fusion blasts the white dwarf apart—and the blast occurs in objects with the same mass and the same composition, strewn throughout the universe. As a result, all of these white dwarf supernovae attain nearly the same maximum energy output, and they all fade away at almost the same rate after they achieve their maximum brightness.

These dual attributes allow SN Ia's to provide astronomers with highly luminous, easily recognizable "standard candles," objects known to achieve the same maximum energy output wherever they appear. Of course, the distance to the supernovae affects their brightnesses as we observe them. Two SN Ia's, seen in two faraway galaxies, will appear to reach the same maximum brightness only if they have the same distance from us. If one has twice the distance of the other, it will attain only one quarter of the other's maximum apparent brightness, because the brightness with which any object appears to us diminishes in proportion to the square of its distance.

Once astronomers learned how to recognize Type Ia supernovae, based on their detailed study of the spectrum of light from each of these objects, they had a golden key with which to unlock the riddle of determining accurate distance. After measuring (through other means) the distances to the closest of the SN Ia's, they could estimate much greater distances to other Type Ia supernovae, simply by comparing the brightnesses of the relatively near and distant objects.

Throughout the 1990s, two teams of supernova specialists, one centered at Harvard and the other at the University of California at Berkeley, refined this technique by finding how to compensate for the small but real differences among the SN Ia's, which the supernovae reveal to us through the details in their spectra. In order to use their newly forged key to unlock the distances to faraway supernovae, the researchers needed a telescope capable of observing distant galaxies with exquisite precision, and they found one in the Hubble Space Telescope, refurbished in 1993 to correct its primary mirror that had been ground to the wrong shape. The supernova experts used ground-based telescopes to discover dozens of SN Ia's in galaxies billions of light-years from the Milky Way. They then arranged for the Hubble Telescope, for which they could obtain only a modest fraction of the total observing time, to study these newfound supernovae in detail.

As the 1990s drew toward a close, the two teams of supernova observers competed keenly to derive a new and expanded "Hubble diagram," the key graph in cosmology that plots galaxies' distances versus the speeds at which the galaxies are moving away from us. Astrophysicists calculate these speeds through their knowledge of the Doppler effect (described in Chapter 13), which changes the colors of the galaxies' light by small amounts that depend on the velocities at which the galaxies are receding from us.

Each galaxy's distance and recession velocity specify a point on the Hubble diagram. For relatively nearby galaxies these points march upward in lockstep, since a galaxy twice as distant from us as another turns out to be receding twice as fast. The direct proportionality between galaxies' distances and recession velocities finds algebraic expression in Hubble's law, the simple equation that describes the basic behavior of the universe: $v = H_o \times d$. Here v stands for recession velocity, d for distance, and H_o is a universal constant, called Hubble's constant, that describes the entire universe at any particular time. Alien observers throughout the universe, studying the cosmos 14 billion years after the big bang, will find galaxies receding at speeds that follow Hubble's law, and all of them will derive the same value for Hubble's constant, though they will probably give it a different name. This assumption of cosmic democracy underlies all of modern cosmology. We cannot prove that the entire cosmos follows this democratic principle. Perhaps, far beyond the farthest horizon of our vision, the cosmos behaves quite differently from what we see. But cosmologists reject this approach, at least for the observable universe. In that case, $v = H_o \times d$ represents universal law.

With time, however, the value of Hubble's constant can and does change. A new and improved Hubble diagram, one that extended to include galaxies many billions of light-years away, will reveal not only the value of today's Hubble constant H_o (embodied in the slope of the line that runs through the points representing galaxies' distances and recession velocities) but also the way in which the universe's current rate of expansion differs from its value billions of years ago. The latter value would be revealed by the details of the upper reaches of the graph, whose points describe the most distant galaxies ever observed. Thus a Hubble diagram extending to distances of many billion light-years would reveal the history of the expansion of the cosmos, embodied in its changing rate of expansion.

In striving for this goal, the astrophysics community struck a mother lode of good fortune in having two competing teams of supernova observers. The supernova results, first announced in February 1998, had an impact so great that no single group could have survived the natural skepticism of cosmologists to the overthrow of their widely accepted models of the universe. Because the two observing teams directed their skepticism primarily at each other, they brilliantly searched for errors in the other team's data or interpretation. When they pronounced themselves satisfied, despite their human prejudices, that their competitors were careful and competent, the cosmological world had little choice but to accept, albeit with some restraint, the news from the frontiers of space.

What was that news? Just that the most distant SN Ia's turned out a bit fainter than expected. This implies that the supernovae are somewhat farther away than they ought to be, which in turn shows that something made the universe expand a bit more rapidly than it should. What provoked this additional expansion? The only culprit that fits the facts is the "dark energy" that lurks in empty space—the energy whose existence corresponds to a non-zero value for the cosmological constant. By measuring the amount by which distant supernovae turned out to be fainter than expected, the two teams of astronomers measured the shape and fate of the universe.

When the two supernova teams achieved consensus, the cosmos turned out to be flat. To understand, we must engage in a bit of rough and tumble in Greek. A universe with a non-zero cosmological constant requires one additional number to describe the cosmos. To the Hubble constant, which we write as H_0 to denote its value at the present time, and to the average density of matter, which alone determines the curvature of space if the cosmological constant is zero, we must now add the density equivalent provided by the dark energy, which, by Einstein's formula $E = mc^2$, must possess the equivalent of mass (m) because it has energy (E). Cosmologists express the densities of matter and dark energy with the symbols Ω_M and Ω_Λ, where Ω (the Greek capital letter omega) stands for the ratio of the cosmic density to the critical density. Ω_M represents the ratio of the average density of all the matter in the universe to the critical density, while Ω_Λ stands for the ratio of the density equivalent provided by the dark energy to the critical density. Here Λ (Greek capital lambda) represents the cosmological constant. In a flat universe, which has zero curvature of space, the sum of Ω_M and Ω_Λ always equals 1, because the total density (of actual matter plus the matter equivalent provided by the dark energy) exactly equals the critical density.

The observations of distant Type Ia supernovae measure the difference between Ω_M and Ω_Λ. Matter tends to slow the expansion of the universe, as gravity pulls everything toward every-

thing else. The greater the density of matter, the more this pull
will slow things down. Dark energy, however, does something
quite different. Unlike pieces of matter, whose mutual attraction
slows the cosmic expansion, dark energy has a strange property: it
tends to make space expand, and thus accelerate the expansion. As
space expands, more dark energy comes into existence, so that the
expanding universe represents the ultimate free lunch. The new
dark energy tends to make the cosmos expand still faster, so the
free lunch grows ever larger as time goes on. The value of Ω_Λ is
a measure of the size of the cosmological constant and gives us
the magnitude of dark energy's expansionist ways. When
astronomers measured the relationship between galaxies' distances and their recession velocities, they found the result of the
contest between gravity's pulling things together and dark
energy's pushing them apart. Their measurements implied that
$\Omega_\Lambda - \Omega_M = 0.46$, plus or minus about 0.03. Since astronomers
had already determined that Ω_M equals approximately 0.25, this
result sets Ω_Λ at about 0.71. Then the sum of Ω_Λ and Ω_M rises to
0.96, near the total predicted by the inflationary model. Recent
new results have sharpened these values and brought this sum
even closer to 1.

Despite the agreement between the two competing groups of
supernova experts, some cosmologists remained cautious. It is not
every day that a scientist abandons a long-held belief, such as the
conviction that the cosmological constant ought to be zero, and
replaces it with a strikingly different one, such as the conclusion
that dark energy fills every cubic centimeter of empty space.
Almost all the skeptics who had followed the ins and outs of cosmological possibilities finally pronounced themselves convinced
after they had digested new observations from a satellite designed
and operated to observe the cosmic background radiation with
unprecedented accuracy. That satellite, the all-important WMAP
described in Chapter 3, began to make useful observations in

2002, and by early 2003 had accumulated sufficient data for cosmologists to make a map of the entire sky, seen in the microwaves that carry most of the cosmic background radiation. Although earlier observations had revealed the basic results to be derived from this map, they had observed only small portions of the sky or shown much less detail. WMAP's whole-sky map provided the capstone to the mapping effort, and has determined, once and for all, the most important features of the cosmic background radiation.

The most striking and significant aspect of this map, as was also true for the balloon-borne observations and for WMAP's predecessor, the COBE (COsmic Background Explorer) satellite, lies in its near featurelessness. No measurable differences in the intensity of the cosmic background radiation arriving from all different directions appear until we reach a precision of about one part in a thousand in our measurements. Even then, the only discernible differences appear as a slightly greater intensity, centered on one particular direction, that matches a corresponding slightly lesser intensity, centered on the opposite direction. These differences arise from our Milky Way galaxy's motion among its neighbor galaxies. The Doppler effect causes us to receive slightly stronger radiation from the direction of this motion, not because the radiation actually is stronger, but because our motion toward the cosmic background radiation (CBR) slightly increases the energies of the photons that we detect.

Once we compensate for the Doppler effect, the cosmic background radiation appears perfectly smooth—until we attain an even higher precision of about one part in a hundred thousand. At that level, tiny deviations from total smoothness appear. They track locations from which the CBR arrives with a bit more, or a bit less, intensity. As previously noted, the differences in intensity mark directions in which matter was either a little hotter and denser, or a little cooler and more rarefied, than the average value

380,000 years after the big bang. The COBE satellite first saw these differences; balloon-borne instruments and South Pole observations improved our measurements; and then the WMAP satellite provided still better precision in surveying the entire sky, allowing cosmologists to construct a detailed map of the intensity of the cosmic background radiation, observed with unprecedented angular resolution of about one degree.

The tiny deviations from smoothness revealed by COBE and WMAP have more than passing interest to cosmologists. First of all, they show the seeds of structure in the universe at the time when the cosmic background radiation ceased to interact with matter. The regions revealed as slightly denser than average at that time had a head start toward further contraction, and have won the competition to acquire the most matter by gravity. Thus the primary result from the new map of the CBR's intensity in different directions is the verification of cosmologists' theories of how the immense differences in density from place to place throughout the cosmos that we see now owe their existence to tiny differences in density that existed a few hundred thousand years after the big bang.

But cosmologists can use their new observations of the cosmic background radiation to discern another, still more basic fact about the cosmos. The details in the map of the CBR's intensity from place to place reveal the curvature of space itself. This amazing result rests on the fact that the curvature of space affects how radiation travels through it. If, for example, space has a positive curvature, then when we observe the cosmic background radiation, we are in much the same situation as an observer at the North Pole who looks along Earth's surface to study radiation produced near the Equator. Because the lines of longitude converge toward the pole, the source of radiation seems to span a smaller angle than it would if space were flat.

To understand how the curvature of space affects the angular size of features in the cosmic background radiation, imagine the time when the radiation finally ceased to interact with matter. At that time, the largest deviations from smoothness that could have existed in the universe had a size that cosmologists can calculate: the age of the universe at that time, multiplied by the speed of light—about 380,000 light-years across. This represents the maximum distance at which particles could have affected one another to produce any irregularities. At greater distances, the "news" from other particles would not yet have arrived, so they cannot be blamed for any deviations from smoothness.

How large an angle would these maximum deviations span on the sky now? That depends on the curvature of space, which we can determine by finding the sum of Ω_M and Ω_Λ. The more closely that sum approaches 1, the more closely the space curvature will approach zero, and the larger will be the angular size that we observe for the maximum deviations from smoothness in the CBR. This space curvature depends only on the sum of the two Ωs, because both types of density make space curve in the same way. Observations of the cosmic background radiation therefore offer a direct measurement of $\Omega_M + \Omega_\Lambda$, in contrast to the supernova observations, which measure the difference between Ω_M and Ω_Λ.

The WMAP data show that the largest deviations from smoothness in the CBR span an angle of about 1 degree, which implies that $\Omega_M + \Omega_\Lambda$ has a value of 1.02, plus or minus 0.02. Thus, within the limits of experimental accuracy, we may conclude that $\Omega_M + \Omega_\Lambda = 1$, and that space is flat. The result from observations of distant SN Ia's may be stated as $\Omega_\Lambda - \Omega_M = 0.46$. If we combine this result with the conclusion that $\Omega_M + \Omega_\Lambda = 1$, we find that $\Omega_M = 0.27$ and $\Omega_\Lambda = 0.73$, with an uncertainty of a few percent in each number. As already noted, these are the astrophysicists' current best estimates for the values of these two key cosmic parameters, which tell us that matter—both ordinary and dark—provides 27 percent of the total energy density in the universe, and dark energy 73 percent. (If we prefer to think of energy's mass equivalent, E/c^2, then dark energy furnishes 73 percent of all the mass.)

Cosmologists have long known that if the universe has a non-zero cosmological constant, the relative influence of matter and dark energy must change significantly as time passes. On the other hand, a flat universe remains flat forever, from its origin in the big bang to the infinite future that awaits us. In a flat universe, the sum of Ω_M and Ω_Λ always equals 1, so if one of these changes, the other must also vary in compensation.

During the cosmic epochs that followed soon after the big bang, the dark energy produced hardly any effect on the universe. So little space existed then, in comparison to the eras that would follow, that Ω_Λ had a value just a bit above zero, while Ω_M was only a tiny bit less than 1. In those bygone ages, the universe behaved in much the same way as a cosmos without a cosmological constant. As time passed, however, Ω_M steadily decreased and Ω_Λ just as steadily increased, keeping their sum constant at 1. Eventually, hundreds of billions of years from now, Ω_M will fall almost all the way to zero and Ω_Λ will rise nearly to unity. Thus, the history of flat space with a non-zero cosmological constant involves a transition from its early years, when the dark energy barely mattered, through the "present" period, when Ω_M and Ω_Λ have roughly equal values, and on into an infinitely long future, when matter will spread so diffusely through space that Ω_M must pursue an infinitely long slide toward zero, even as the sum of the two Ωs remains equal to 1.

Observational deduction of how much mass exists in galaxy clusters now gives Ω_M a value of about 0.25, while the observations of the CBR and distant supernovae imply a value close to 0.27. Within the limits of experimental accuracy, these two values coincide. If the universe in which we live does have a non-zero cosmological constant, and if that constant is responsible (along with the matter) for producing the flat universe that the inflationary model predicts, then the cosmological constant must have a value that makes Ω_Λ equal to a bit more than 0.7, two and a half times the value of Ω_M. In other words, Ω_Λ must now do most of the work in making $(\Omega_M + \Omega_\Lambda)$ equal to 1. This means that we have already passed through the cosmic era when matter and the cosmological constant contributed the same amount (with each of them equal to 0.5) toward maintaining the flatness of space.

Within less than a decade, the double-barreled blast from the Type Ia supernovae and the cosmic background radiation has changed the status of dark energy from a far-out idea that Einstein once toyed with to a cosmic fact of life. Unless a host of observations eventually prove to be misinterpreted, inaccurate, or just plain wrong, we must accept the result that the universe will never contract or recycle itself. Instead, the future seems bleak: a hundred billion years from now, when most stars will have burnt themselves out, all but the closest galaxies will have vanished across our horizon of visibility.

By then, the Milky Way will have coalesced with its nearest neighbors, creating one giant galaxy in the literal middle of nowhere. Our night sky will contain orbiting stars, (dead and alive) and nothing else, leaving future astrophysicists a cruel universe. With no galaxies to track the cosmic expansion, they will erroneously conclude, as did Einstein, that we live in a static universe. The cosmological constant and its dark energy will have evolved the universe to a point where they cannot be measured or even dreamt of.

Enjoy cosmology while you can.

CHAPTER 6

One Universe or Many?

The discovery that we live in an accelerating universe, with an ever-increasing rate of expansion, rocked the world of cosmology early in 1998, with the first announcement of the supernova observations that point to this acceleration. Now that the accelerating universe has received confirmation from detailed observations of the cosmic background radiation, and now that cosmologists have had several years to wrestle with the implications of an accelerating cosmic expansion, two great questions have emerged to bedevil their days and brighten their dreams: What makes the universe accelerate? And why does that acceleration have the particular value that now characterizes the cosmos?

The simple answer to the first question assigns all responsibility for the acceleration to the existence of dark energy, or, equivalently, to a non-zero cosmological constant. The amount of acceleration depends directly on the amount of dark energy per cubic centimeter: More energy implies greater acceleration.

Thus, if cosmologists could only explain where the dark energy comes from, and why it exists in the amount that they find today, they could claim to have uncovered a fundamental secret of the universe—the explanation for the cosmic "free lunch," the energy in empty space that continuously drives the cosmos toward an eternal, ever more rapid expansion and a far future of enormous amounts of space, correspondingly enormous amounts of dark energy, and almost no matter per cubic light-year.

What makes dark energy? From the deep realms of particle physics, cosmologists can produce an answer: The dark energy arises from events that must occur in empty space, if we trust what we have learned from the quantum theory of matter and energy. All of particle physics rests on this theory, which has been verified so often and so exactly in the submicroscopic realm that almost all physicists accept it as correct. An integral part of quantum theory implies that what we call empty space in fact buzzes with "virtual particles," which wink into and out of existence so rapidly that we can never pin them down directly, but can only observe their effects. The continual appearance and disappearance of these virtual particles, called the "quantum fluctuations of the vacuum" by those who like a good physics phrase, gives energy to empty space. Furthermore, particle physicists can, without much difficulty, calculate the amount of energy that resides in every cubic centimeter of the vacuum. The straightforward application of quantum theory to what we call a vacuum predicts that quantum fluctuations must create dark energy. When we tell the story from this perspective, the great question about dark energy seems to be, Why did cosmologists take so long to recognize that this energy must exist?

Unfortunately, the details of the actual situation turn this question into, How did particle physicists go so far wrong? Calculations of the amount of dark energy that lurks in every cubic centimeter produce a value about 120 powers of ten greater than the value

that cosmologists have found from observations of supernovae
and the cosmic background radiation. In far-out astronomical situations, calculations that prove correct to within a single factor of
10 are often judged at least temporarily acceptable, but a factor of
10^{120} cannot be swept under the rug, even by physics Pollyannas.
If real empty space contained dark energy in anything like the
amounts proposed by particle physics, the universe would have
long since puffed itself into so large a volume that our heads could
never have begun to spin, since a tiny fraction of a second would
have sufficed to spread matter out to unimaginable rarefaction.
Theory and observation agree that empty space ought to contain
dark energy, but they disagree by a trillion to the tenth power
about the amount of that energy. No earthly analogy, nor even a
cosmic one, can illustrate this discrepancy accurately. The distance to the farthest galaxy that we know exceeds the size of a
proton by a factor of 10^{40}. Even this enormous number is only the
cube root of the factor by which theory and observation currently
diverge concerning the value of the cosmological constant.

Particle physicists and cosmologists have long known that
quantum theory predicts an unacceptably large value for the dark
energy, but in the days when the cosmological constant was
thought to be zero, they hoped to discover some explanation that
would, in effect, cancel positive with negative terms in the theory
and thereby finesse the problem out of existence. A similar cancellation once solved the problem of how much energy virtual
particles contribute to the particles that we do observe. Now that
the cosmological constant turns out to be non-zero, the hopes of
finding such a cancellation seem dimmer. If the cancellation does
exist, it must somehow remove almost all of the mammoth theoretical value we have today. For now, lacking any good explanation for the size of the cosmological constant, cosmologists must
continue to collaborate with particle physicists as they seek to
reconcile theories of how the cosmos generates dark energy with

the value observed for the amount of dark energy per cubic centimeter.

Some of the finest minds engaged in cosmology and particle physics have directed much of their energy toward explaining this observational value, with no success at all. This provokes fire, and sometimes ire, among theorists, in part because they know that a Nobel Prize—not to mention the immense joy of discovery—awaits those who can explain what nature has done to make space as we find it. But another issue stokes intense controversy as it cries out for explanation: Why does the amount of dark energy, as measured by its mass equivalent, roughly equal the amount of energy provided by all the matter in the universe?

We can recast this question in terms of the two Ωs that we use to measure the density of matter and the density equivalent of dark energy: Why do Ω_M and Ω_Λ roughly equal one another, rather than one being enormously larger than the other? During the first billion years after the big bang, Ω_M was almost precisely equal to 1, while Ω_Λ was essentially zero. In those years, Ω_M was first millions, then thousands, and afterward hundreds of times greater than Ω_Λ. Today, with $\Omega_M = 0.27$ and $\Omega_\Lambda = 0.73$, the two values are roughly equal, though Ω_Λ is already notably larger than Ω_M. In the far future, more than 50 billion years from now, Ω_Λ will be first hundreds, then thousands, after that millions, and still later billions of times greater than Ω_M. Only during the cosmic era from about 3 billion to 50 billion years after the big bang do the two quantities match one other even approximately.

To the easygoing mind, the interval between 3 billion and 50 billion years embraces quite a long period of time. So what's the problem? From an astronomical viewpoint, this stretch of time amounts to nearly nothing. Astronomers often take a logarithmic approach to time, dividing it into intervals that increase by factors of 10. First the cosmos had some age; then it grew ten times older; then ten times older than that; and so on toward infinite time, which requires an infinite number of ten-times jumps. Suppose that we start counting time at the earliest moment after the big bang that has any significance in quantum theory, 10^{-43} second after the big bang. Since each year contains about 30 million (3 x 10^7) seconds, we need about 60 factors of 10 to pass from 10^{-43} second to 3 billion years after the big bang. In contrast, we require only a bit more than a single factor of 10 to stroll from 3 billion to 50 billion years, the only period when Ω_M and Ω_Λ are roughly equal. After that, an infinite number of ten-times factors opens the way to the infinite future. From this logarithmic perspective, only a vanishingly small probability exists that we should find ourselves living in a cosmic situation for which Ω_M and Ω_Λ have even vaguely similar values. Michael Turner, a leading American cosmologist, has termed this conundrum—the question of why we find ourselves alive at a time when Ω_M and Ω_Λ are approximately equal—the "Nancy Kerrigan problem" in honor of the Olympic figure skater, who asked, after enduring an assault by her rival's boyfriend, "Why me? Why now?"

Despite their inability to calculate a cosmological constant whose value comes anywhere close to the measure one, cosmologists do have an answer to the Kerrigan problem, but they differ sharply on its significance and implications. Some embrace it; some accept it only reluctantly; some dance around it; and some despise it. This explanation links the value of the cosmological constant to the fact that we are here, alive on a planet that orbits an average star in an average galaxy. Because we exist, this argument runs, the parameters that describe the cosmos, and in particular the value of the cosmological constant, must have values that allow us to exist.

Consider, for example, what would happen in a universe with a cosmological constant much larger than its actual value. A much greater amount of dark energy would make Ω_Λ rise far above Ω_M, not after about 50 billion years but after only a few million years.

By this time, in a cosmos dominated by the accelerating effects of dark energy, matter would spread so rapidly apart that no galaxies, stars, or planets could form. If we assume that the stretch of time from the first formation of clumps of matter to the origin and development of life covers at least 1 billion years, we can conclude that our existence limits the cosmological constant to a value between zero and a few times its actual value, while ruling out of play the infinite range of higher values.

This argument gains more traction if we assume, as do many cosmologists, that everything we call the universe belongs to a much larger "multiverse," which contains an infinite number of universes, none of which interact with any other: in the multiverse concept, the entire state of affairs embeds in higher dimensions, so space in our universe remains completely inaccessible to any other universe, and vice versa. This lack of even theoretically possible interactions puts the multiverse theory into the category of apparently nontestable, and therefore nonverifiable, hypotheses—at least until wiser minds find ways to test the multiverse model. In the multiverse, new universes are born at completely random times, capable of swelling up by inflation into enormous volumes of space, and of doing so without interfering in the least with the infinite number of other universes.

In the multiverse, each new universe springs into existence with its own physical laws and its own set of cosmic parameters, including the rules that determine the size of the cosmological constant. Many of these other universes have cosmological constants enormously larger than ours, and quickly accelerate themselves into situations of near-zero density, no good for life. Only a tiny, perhaps an infinitesimal fraction of all the universes in the multiverse offer conditions that allow life to exist, because only this fraction have parameters that permit matter to organize itself into galaxies, stars, and planets, and for those objects to last for billions of years.

Cosmologists call this approach to explaining the value of the cosmological constant the anthropic principle, though the anthropic approach probably offers a better name. This approach toward explaining a crucial question in cosmology has one great appeal: people love it or hate it, but rarely feel neutral about it. Like many intriguing ideas, the anthropic approach can be bent to favor, or at least seem to favor, various theological and teleological mental edifices. Some religious fundamentalists find that the anthropic approach supports their beliefs because it implies a central role for humanity: without someone to observe it, the cosmos—at least the cosmos that we know—would not, could not, be here; hence a higher power must have made things just right for us. An opponent of this conclusion would note this is not really what the anthropic approach implies; on a theological level, this argument for the existence of God implies surely the most wasteful creator one might imagine, who makes countless universes in order that in a tiny sector of just one of these, life might arise. Why not skip the middleman and follow older creation myths that center on humanity?

On the other hand, if you choose to see God in everything, as Spinoza did, you cannot help but admire a multiverse that effloresces universes without end. Like most news from the frontier of science, the concept of a multiverse, and the anthropic approach, can be easily bent in different directions to serve the needs of particular belief systems. As things stand, many cosmologists find the multiverse quite enough to swallow without connecting it to any system of beliefs. Stephen Hawking, who (like Isaac Newton before him) holds the Lucasian chair in astronomy at Cambridge University, judges the anthropic approach an excellent resolution of the Kerrigan problem. Stephen Weinberg, who won the Nobel Prize for his insights into particle physics, dislikes this approach but pronounces himself in favor, at least for the time being, because no other reasonable solution has appeared.

History may eventually show that for now, cosmologists are concentrating on the wrong problem—wrong in the sense that we don't yet understand enough to attack it properly. Weinberg likes the analogy with Johannes Kepler's attempt to explain why the Sun has six planets (as astronomers then believed), and why they move in the orbits that they do. Four hundred years after Kepler, astronomers still know far too little about the origin of planets to explain the precise number and orbits of the Sun's family. We do know that Kepler's hypothesis, which proposed that the spacing of the planets' orbits around the Sun allows one of the five perfect solids to fit exactly between each pair of adjoining orbits, has no validity whatsoever, because the solids do not fit particularly well, and (even more important) because we have no good reason to explain why the planets' orbits should obey such a rule. Later generations may regard today's cosmologists as latter-day Keplers, struggling valiantly to explain what remains inexplicable by today's understanding of the universe.

Not everyone favors the anthropic approach. Some cosmologists attack it as defeatist, ahistorical (since this approach contradicts numerous examples of the success of physics in explaining, sooner or later, a host of once mysterious phenomena), and dangerous, because the anthropic approach smacks of intelligent-design arguments. Furthermore, many cosmologists find unacceptable, as grounds for a theory of the universe, the assumption that we live in a multiverse that contains a multitude of universes with which we can never interact, even in theory.

The debate over the anthropic principle highlights the skepticism that underlies the scientific approach to understanding the cosmos. A theory that appeals to one scientist, typically the one who thought it up, may seem ridiculous, or just plain wrong, to another. Both know that theories survive and thrive when other scientists find them best at explaining most of the observational data. (As a famous scientist once remarked, Beware of a theory that explains *all* the data—some of it will quite likely turn out to be wrong.)

The future may not produce a quick resolution to this debate, but it will surely bring forth other attempts to explain what we see in the universe. For example, Paul Steinhardt of Princeton University, who could use some tutoring in creating catchy names, has produced a theoretical "ekpyrotic model" of the cosmos in collaboration with Neil Turok of Cambridge University. Motivated by the section of particle physics called string theory, Steinhardt envisions a universe with eleven dimensions, most of which are "compactified," more or less rolled up like a sock, so that they occupy only infinitesimal amounts of space. But some of the additional dimensions have real size and significance, except that we can't perceive them because we remain locked into our familiar four. If you pretend that all of space in our universe fills an infinite thin sheet (this model reduces the three dimensions of space to two), you can imagine another, parallel sheet, and then picture the two sheets approaching and colliding. The collision produces the big bang, and as the sheets rebound from one another, each sheet's history proceeds along familiar lines, giving birth to galaxies and stars. Eventually, the two sheets cease to separate and start to approach one another again, producing another collision and another big bang in each sheet. The universe thus has a cyclical history, repeating itself, at least in its broadest outlines, at intervals of hundreds of billions of years. Since "ekpyrosis" means "conflagration" in Greek (recall the more familiar word "pyromaniac"), the "ekpyrotic universe" reminds all those with Greek at the tips of their brains of the great fire that gave birth to the cosmos that we know.

This ekpyrotic model of the universe has emotional and intellectual appeal, though not enough to win the hearts and minds of many of Steinhardt's fellow cosmologists. Not yet, anyhow. Something vaguely like the ekpyrotic model, if not this model itself, may someday offer the breakthrough that cosmologists now await in their attempts to explain the dark energy. Even those who favor the anthropic approach would hardly dig in their heels to resist a new theory that could provide a good explanation for the cosmological constant without invoking an infinite number of universes, of which ours happens to be one of the lucky ones. As one of R. Crumb's cartoon characters once said: "What a wonderful, wacky world we live in! Wooey!"

Part II

The Origin of
Galaxies and
Cosmic Structure

CHAPTER 7

Discovering Galaxies

T wo and a half centuries ago, shortly before the English
astronomer Sir William Herschel built the world's first seriously large telescope, the known universe consisted of little
more than the stars, the Sun and Moon, the planets, a few moons
of Jupiter and Saturn, some fuzzy objects, and the galaxy that
forms a milky band across the night sky. Indeed, the word
"galaxy" derives from the Greek *galaktos*, or "milk." The sky also
held the fuzzy objects, scientifically named nebulae after the
Latin word for clouds—objects of indeterminate shape such as
the Crab nebula in the constellation Taurus, and the Andromeda
nebula, which appears to live among the stars of the constellation
Andromeda.

Herschel's telescope had a mirror forty-eight inches across, an
unprecedented size for 1789, the year it was built. A complex
array of trusses to support and point this telescope made it an
ungainly instrument, but when he aimed it at the heavens, Herschel could readily see the countless stars that compose the Milky

Way. Using his forty-eight-incher, as well as a smaller, more nimble telescope, Herschel and his sister Caroline compiled the first extensive "deep sky" catalogue of northern nebulae. Sir John—Herschel's son—continued this family tradition, adding to his father's and aunt's list of northern objects and, during an extended stay at the Cape of Good Hope at the southern tip of Africa, cataloguing some 1,700 fuzzy objects visible from the Southern Hemisphere. In 1864, Sir John produced a synthesis of the known deep sky objects: *A General Catalogue of Nebulae and Clusters of Stars*, which included more than five thousand entries.

In spite of that large body of data, nobody at the time knew the true identity of the nebulae, their distances from Earth, or the differences among them. Nevertheless, the 1864 catalogue made it possible to classify the nebulae morphologically—that is, according to their shapes. In the "we call 'em as we see 'em" tradition of baseball umpires (who came into their own just about the time that Herschel's *General Catalogue* was published), astronomers named the spiral-shaped nebulae "spiral nebulae," those with a vaguely elliptical shape "elliptical nebulae," and the various irregularly shaped nebulae—neither spiral nor elliptical—"irregular nebulae." Finally, they called the nebulae that looked small and round, like a telescopic image of a planet, "planetary nebulae," an act that has permanently confused newcomers to astronomy.

For most of its history, astronomy has remained plainspoken, using descriptive methods of inquiry that greatly resembled those used in botany. Using their lengthening compendia of stars and fuzzy things, astronomers searched for patterns and sorted objects according to them. Quite a sensible step, too. Most people, beginning in childhood, arrange things according to appearance and shape without even being told to do so. But this approach can carry you only so far. The Herschels always assumed, because many of their fuzzy objects span a patch of about the same size on the night sky, that all the nebulae lay at about the same distance from Earth. So to them it was simply good, evenhanded science to subject all the nebulae to the same rules of sorting.

Trouble is, the assumption that all nebulae lay at similar distances turned out to be badly mistaken. Nature can be elusive, even devious. Some of the nebulae classified by the Herschels are no farther away than the stars, so they are relatively small (if a trillion miles across can be called "relatively small"). Others turned out to be much more distant, so they must be much larger than the fuzzy objects relatively close to us if they are to appear the same size on the sky.

The take-home lesson is that at some point you've got to stop fixating on what something looks like and start asking what it is. Fortunately, by the late nineteenth century, advances in science and technology had empowered astronomers to do just that, to move beyond merely classifying the contents of the universe. That shift led to the birth of astrophysics, the useful application of the laws of physics to astronomical situations.

During the same era when Sir John Herschel published his vast catalogue of nebulae, a new scientific instrument, the spectroscope, joined the search for nebulae. The sole job of a spectroscope is to break light into a rainbow of its component colors. Those colors, and features embedded within them, reveal not only fine details about the chemical composition of the light source but also, because of a phenomenon called the Doppler effect, the motion of the light source toward or away from Earth.

Spectroscopy revealed something remarkable: the spiral nebulae, which happen to predominate outside the swath of the Milky Way, are nearly all moving away from Earth, and at extremely high speeds. In contrast, all the planetary nebulae, as well as most irregular nebulae, are traveling at relatively low speeds—some toward us and some away from us. Had some catastrophic explosion taken place in the center of the Milky Way, booting out only the spiral nebulae? If so, why weren't any of them falling back? Were we catching the catastrophe at a special time? In spite of advances in photography that brought forth faster emulsions, enabling astronomers to measure the spectra of ever dimmer nebulae, the exodus continued and these questions remained unanswered.

Most advances in astronomy, as in other sciences, have been driven by the introduction of better technology. As the 1920s opened, another key instrument appeared on the scene: the formidable 100-inch Hooker Telescope at the Mount Wilson Observatory near Pasadena, California. In 1923, the American astronomer Edwin P. Hubble used this telescope—the largest in the world at that time—to find a special breed of star, a Cepheid variable star, in the Andromeda nebula. Variable stars of any type vary in brightness according to well-known patterns; Cepheid variables, named for the prototype of the class, a star in the constellation Cepheus, are all extremely luminous and therefore visible over vast distances. Because their brightness varies in recognizable cycles, patience and persistence will yield an increasing number to the careful observer. Hubble had found a few of these Cepheid variables within the Milky Way and estimated their distances; yet, to his astonishment, the Cepheid he found in Andromeda was much dimmer than any of those.

The most likely explanation for this dimness was that the new Cepheid variable, and the Andromeda nebula in which it lives, sits at a distance much greater than those to the Cepheids in the Milky Way. Hubble realized that this placed the Andromeda nebula at so great a distance that it could not possibly lie among the stars in the constellation Andromeda, nor anywhere within the Milky Way—and could not have been kicked out, along with all its spiral sisters, during a catastrophe milk spill.

The implications were breathtaking. Hubble's discovery showed that spiral nebulae are entire systems of stars in their own right, as huge and as packed with stars as our own Milky Way. In the phrase of the philosopher Immanuel Kant, Hubble had demonstrated that "island universes" by the dozens must lie outside our own star system, for the object in Andromeda merely led the list of well-known spiral nebulae. The Andromeda nebula was, in fact, the Andromeda *galaxy*.

By 1936, enough island universes had been identified and photographed through the Hooker and other large telescopes that Hubble, too, decided to try his hand at morphology. His analysis of galaxy types rested upon the untested assumption that variations from one shape to another among galaxies signify evolutionary steps from birth to death. In his 1936 book *Realm of the Nebulae,* Hubble classified galaxies by placing the different types along a diagram shaped like a musical tuning fork, whose handle represents the elliptical galaxies, with rounded ellipticals at the far end of the handle and flattened ellipticals near the point where the two tines join. Along one tine lie the ordinary spiral galaxies: those nearest the handle have their spiral arms wound extremely tightly, while those toward the tine's end have increasingly loosely wound spiral arms. Along the other tine are spiral galaxies whose central region displays a straight "bar," but are otherwise similar to ordinary spirals.

Hubble imagined that galaxies start their lives as round ellipticals and become flatter and flatter as they continue to take shape, ultimately revealing a spiral structure that slowly unfurls with the passage of time. Brilliant. Beautiful. Even elegant. But just plain wrong. Not only were entire classes of irregular galaxies omitted from this scheme, but astrophysicists would later learn that the oldest stars in every galaxy were about the same age, implying that all the galaxies were born during a single era in the history of the universe.

For three decades (with some research opportunities lost because of World War II), astronomers observed and catalogued galaxies in accordance with Hubble's tuning-fork diagram as ellipticals, spirals, and barred spirals, with irregulars a minority subset, completely off the chart because of their strange shapes. Of elliptical galaxies one might say, as Ronald Reagan did about California's redwoods, that when you've seen one, you've seen them all. Elliptical galaxies resemble one another in possessing neither the spiral-arm patterns that characterize spirals and barred spirals, nor the giant clouds of interstellar gas and dust that give birth to new stars. In these galaxies, star formation ended many billion years ago, leaving behind spherical or ellipsoidal groups of stars. The largest elliptical galaxies, like the largest spirals, each contain many hundred billion stars—perhaps even a trillion or more—and have diameters close to a hundred thousand light-years. With the exception of professional astronomers, no one has ever sighed over the fantastic patterns and complex star formation histories of an elliptical galaxy for the excellent reason that, at least in comparison with spirals, ellipticals have simple shapes and straightforward star formation: they all turned gas and dust into stars until they could do so no more.

Happily, spirals and barred spirals furnish the visual excitement so lacking in ellipticals. The most deeply resonant of all the galaxy images that we may ever see, a view of the entire Milky Way taken from outside it, will stir our hearts and minds, just as soon as we manage to send a camera several hundred thousand light-years above or below the central plane of our galaxy. Today, when our most far-flung space probes have traveled a billionth of that distance, this goal may seem unattainable, and indeed even a probe that could reach nearly the speed of light would require a long wait—far longer than the current span of recorded history—to yield the desired result. For the time being, astronomers must continue to map the Milky Way from inside, sketching the galactic forest by delineating its stellar and nebular trees. These efforts reveal that our galaxy closely resembles our closest large neighbor, the great spiral galaxy in Andromeda. Conveniently located about 2.4 million light-years away, the Andromeda galaxy has provided a wealth of information about the basic structural patterns of spiral galaxies, as well as about different types of stars and their evolution. Because all of the Andromeda galaxy's stars have the same distance from us (plus or minus a few percent), astronomers know that the stars' brightnesses correlate directly with their luminosities, that is, with the amounts of energy they emit each second. This fact, denied to astronomers when they study objects in the Milky Way but applicable to every galaxy beyond our own, has allowed them to draw key conclusions about stellar evolution with greater ease than would be true for stars in the Milky Way. Two elliptical satellite galaxies that orbit the Andromeda galaxy, each containing a few percent of the number of stars in the main galaxy, have likewise furnished important information about the lives of stars, and the overall galactic struc-ture of elliptical galaxies. On a clear night far from city lights, a keen-eyed observer who knows where to look can spot the fuzzy outline of the Andromeda galaxy—the most distant object visible to the unaided eye, shining with light that left on its journey while our ancestors roamed the gorges of Africa in search of roots and berries.

Like the Milky Way, the Andromeda galaxy lies midway along one tine of Hubble's tuning-fork diagram, because its spiral arms are neither particularly tightly nor loosely wound. If galaxies were animals in a zoo, there would be one cage devoted to ellipti-cals but several animal houses for the glorious spirals. To study a Hubble Telescope image of one of these beasts, typically (for the closer ones) seen from 10 or 20 million light-years, is to enter a world of sight so rich in possibility, so deep in separation from life on Earth, so complex in structure, that the unprepared mind may reel, or may provide a defense by reminding its owner that none of this can thin the thighs or heal the fractured bone.

Irregulars, the orphans of the galactic class system, comprise about 10 percent of all galaxies, with the rest split between spirals and ellipticals, strongly favoring spirals. In contrast to ellipticals, irregular galaxies typically contain a higher proportion of gas and dust than spirals do, and offer the liveliest sites of ongoing star formation. The Milky Way has two large satellite galaxies, both irregular, confusingly named the Magellanic Clouds because the first white men to notice them, sailors on Magellan's circumnavigation of Earth in 1520, thought at first they were seeing wisps of clouds in the sky. This honor fell to Magellan's expedition because the Magellanic Clouds lie so close to the south celestial pole (the point directly above Earth's South Pole) that they never rise above the horizon for observers in the most populated Northern latitudes, including those in Europe and most of the United States. Each of the Magellanic Clouds contains many billion stars, though not the hundreds of billions that characterize the Milky Way and other large galaxies, and display immense star-forming regions, most notably the "Tarantula nebula" of the Large Magellanic Cloud. This galaxy also has the honor of having revealed the closest and brightest supernova to appear during the past three centuries, Supernova 1987A, which must have actually exploded about 160,000 B.C. for its light to reach Earth in 1987.

Until the 1960s, astronomers were content to classify almost all galaxies as spiral, barred spiral, elliptical, or irregular. They had right on their side, since more than 99 percent of all galaxies fit one of these classes. (With one galactic class called "irregular," this result might seem to be a slam dunk.) But during that fine decade, an American astronomer named Halton Arp became the champion of galaxies that did not fit the simple classification scheme of the Hubble tuning-fork diagram plus irregulars. In the spirit of "Give me your tired, your poor, your huddled masses," Arp used the world's largest telescope, the 200-inch Hale Telescope at the Palomar Observatory near San Diego, California, to photograph 338 extremely disturbed-looking systems. Arp's *Atlas of Peculiar Galaxies*, published in 1966, became a veritable treasure chest of research opportunities on what can go bad in the universe. Although "peculiar galaxies"—defined as galaxies with such strange shapes that even "irregular" fails to do them justice— form only a tiny minority of all galaxies, they carry important information about what can happen to galaxies gone wrong. It turns out, for example, that many embarrassingly peculiar galaxies in Arp's atlas are the merged remnants of two once-separate galaxies that have collided. This means that those "peculiar" galaxies are not different kinds of galaxies at all, any more than a wrecked Lexus is a new kind of car.

To track how such a collision unfolds, you need a lot more than pencil and paper, because every star in both galactic systems has its own gravity, which simultaneously affects all the other stars in the two systems. What you need, in short, is a computer. Galaxy collisions are stately dramas, taking hundreds of millions of years from beginning to end. Using a computer simulation, you can start, and pause as you like, a collision of two galaxies, taking snapshots after 10 million years, 50 million years, 100 million years. At each time things look different. And when you step into Arp's atlas—batta-bing—here's an early stage of a collision, and there's a late stage. Here's a glancing blow, and there's a head-on collision.

Although the first computer simulations were done in the early 1960s (and although the Swedish astronomer Erik Holmberg made a clever attempt during the 1940s to recreate a galaxy collision on a tabletop by using light as an analogue to gravity), it wasn't until 1972 that Alar and Juri Toomre, brothers who both teach at MIT, generated the first compelling portrait of a "deliberately simple-minded" collision between two spiral galaxies. The Toomres' model revealed that tidal forces—differences in gravity from place to place—actually rip the galaxies apart. As one galaxy nears the other, the gravitational force rapidly grows stronger at the leading edges of the collision, stretching and warping both galaxies as they pass by or through each other. That stretching and warping accounts for most of what's peculiar in Arp's atlas of peculiar galaxies.

How else can computer simulations help us to understand galaxies? Hubble's tuning fork distinguishes "normal" spiral galaxies from spirals that show a dense bar of stars across their centers. Simulations show that this bar could be a transitory feature, not the distinguishing mark of a different galactic species. Contemporary observers of barred spirals might simply be catching such galaxies during a phase that will disappear in 100 million years or so. But since we can't hang around long enough to watch the bar disappear in real life, we have to watch it come and go on a computer, where a billion years can unfold in a matter of minutes.

Arp's peculiar galaxies proved to be the tip of an iceberg, a strange world of not-exactly-galaxies whose outlines astronomers began to discern during the 1960s and came to understand a few decades later. Before we can appreciate this emergent galactic zoo, we must resume the story of cosmic evolution where we left it. We must examine the origin of all galaxies—normal, nearly normal, irregular, peculiar, and knock-your-socks-off exotic—to see how they were born, and how the luck of the draw has left us in our relatively calm location in space, adrift in the suburbs of a giant spiral galaxy, some 30,000 light-years from its center and twenty-thousands of light-years from its diffuse outer edge.

Thanks to the general order of things in a spiral galaxy, first imposed on the gas clouds that later gave birth to stars, our Sun moves in a nearly circular orbit around the center of the Milky Way, taking 240 million years (sometimes called a "cosmic year") for each trip. Today, twenty orbits after its birth, the Sun should be good for another twenty or so before calling it quits. Meanwhile, let's have a look at where galaxies came from.

CHAPTER 8

The Origin of Structure

W hen we examine the history of matter in the universe, looking back through 14 billion years of time as best we can, we quickly encounter a single trend that cries out for explanation. Throughout the cosmos, matter has consistently organized itself into structures. From its nearly perfectly smooth distribution soon after the big bang, matter has clumped itself together on all size scales, to produce giant clusters and superclusters of galaxies, as well as the individual galaxies within those clusters, the stars that congregate by the billions in every galaxy, and quite possibly much smaller objects—planets, their satellites, asteroids, and comets—that orbit many if not most of those stars.

To understand the origin of the objects that now compose the visible universe, we must focus on the mechanisms that turned the universe's formerly diffuse matter into highly structured components. A complete description of how structures emerged in the cosmos requires that we meld two aspects of reality whose combination now eludes us. As seen in earlier chapters, we must perceive how quantum mechanics, which describes the behavior of molecules, atoms, and the particles that form them, fits with general relativity theory, which describes how extremely large amounts of matter and space affect one another.

Attempts to create a single theory that would unite our knowledge of the sub-atomically small and the astronomically large began with Albert Einstein. They have continued, with relatively little success, right up to the present time and will endure into an uncertain future, until they achieve "grand unification." Among all the unknowns that irk them, modern cosmologists feel most acutely the lack of a theory that triumphantly blends quantum mechanics with general relativity. Meanwhile, these seemingly immiscible branches of physics—the science of the small and the science of the large—care not a whit for our ignorance; instead, they co-exist with remarkable success inside the same universe, mocking our attempts to understand them as a coherent whole. A galaxy with 100 billion stars apparently pays no particular attention to the physics of the atoms and molecules that compose its star systems and gas clouds. Neither do the even larger agglomerations of matter we call galaxy clusters and superclusters, themselves containing hundreds, sometimes thousands of galaxies. But these largest structures in the universe nonetheless owe their very existence to immeasurably small quantum fluctuations within the primeval cosmos. To understand how these structures arose, we must do the best we can in our current state of ignorance, passing from the minuscule domains governed by quantum mechanics, which hold the key to the origin of structure, to those so large that quantum mechanics plays no role, and matter obeys the laws laid down by general relativity.

To this end, we must seek to explain the structure-laden universe that we see today as arising from a nearly featureless cosmos soon after the big bang. Any attempt to explain the origin of structure must also account for the cosmos in its present state. Even this modest task has confounded astronomers and cosmologists with a series of false starts and errors, from which we have now (so we may fervently hope) removed ourselves to walk in the bright light of a correct description of the universe.

Throughout most of modern cosmology's history, astrophysicists have assumed that the distribution of matter in the universe can be described as both homogenous and isotropic. In a homogenous universe, every location looks similar to every other location, like the contents of a glass of homogenized milk. An isotropic universe is one that looks the same in every direction from any given point in space and time. These two descriptions may seem the same, but they are not. For example, the lines of longitude on Earth are not homogeneous, because they are farther apart in some regions and closer together in others; they are isotropic in just two places, the North and South poles, where all lines of longitude converge. If you stand at either the "top" or "bottom" of the world, the longitude grid will look the same to you, no matter how far to the left or the right you turn your head. In a more physical example, imagine yourself atop a perfect, cone-shaped mountain, and imagine that this mountain is the only thing in the world. Then every view of Earth's surface from that perch would look the same. The same would be true if you happened to live in the center of an archery target, or if you were a spider at the center of its perfectly spun web. In each of these cases, your view will be isotropic, but decidedly not homogenous.

An example of a homogeneous but non-isotropic pattern appears in a wall of identical rectangular bricks, laid in a bricklayer's traditional, overlapping manner. On the scale of several adjoining bricks and their mortar, the wall will be the same everywhere—bricks—but different lines of sight along the wall will intersect the mortar differently, destroying any claim to isotropy.

Intriguingly (for those who love a certain kind of intrigue), mathematical analysis tells us that space will turn out to be homogeneous only if it is everywhere isotropic. Another formal theorem of mathematics tells us that if space is isotropic in just three places, then space must be isotropic everywhere. Yet some of us shun mathematics as uninteresting and unproductive!

Although cosmologists were aesthetically motivated for assuming the homogeneity and isotropy of the distribution of matter in space, they have come to believe in this assumption enough to establish it as a fundamental cosmological principle. We might also call this the principle of mediocrity: Why should one part of the universe be any more interesting than another? On the smallest scales of size and distance, we easily recognize this assertion to be false. We live on a solid planet with an average density of matter close to 5.5 grams per cubic centimeter (in Americanese, that's about 340 pounds per cubic foot). Our Sun, a typical star, has an average density of about 1.4 grams per cubic centimeter. The interplanetary spaces between the two, however, have a significantly smaller average density—smaller by a factor of about 1 billion trillion. Intergalactic space, which accounts for most of the volume of the universe, contains less than one atom in every ten cubic meters. Here the average density falls below the density of interplanetary space by another factor of 1 billion—enough to make the mind feel good about the occasional accusation of being dense.

As astrophysicists expanded their horizons, they saw clearly that a galaxy such as our Milky Way consists of stars that float through nearly empty interstellar space. The galaxies likewise group into clusters that violate the assumption of homogeneity and isotropy. The hope remained, however, that as astrophysicists charted visible matter on the largest scales, they would find that galaxy clusters have a homogenous and isotropic distribution. For homogeneity and isotropy to exist within a particular region of space, it must be large enough that no structures (or lack of structures) sit uniquely within it. If you take a melon-ball sample of such a region, the requirements of homogeneity and isotropy imply that the region's overall properties must be similar in every way to the average properties of any other scoop with the same size. What an embarrassment it would be if the left half of the universe looked different from its right half.

How large a region must we examine to find a homogeneous and isotropic universe? Our planet Earth has a diameter of 0.04 light-seconds. Neptune's orbit spans 8 light-hours. The stars of the Milky Way galaxy delineate a broad, flat disk about 100,000 light-years across. And the Virgo supercluster of galaxies, to which the Milky Way belongs, extends some 60 million light-years. So the coveted volume that can give us homogeneity and isotropy must be larger than the Virgo supercluster. When astrophysicists made surveys of the galaxies' distribution in space, they discovered that even on these scales of size, as large as 100 million light-years, the cosmos reveals enormous, comparatively empty gaps, bounded by galaxies that have arranged themselves into intersecting sheets and filaments. Far from resembling a teeming, homogenous anthill, the distribution of galaxies on this scale resembles a loofah sponge.

Finally, however, astrophysicists made still larger maps, and found their treasured homogeneity and isotropy. Turns out, the contents of a 300-million-light-year scoop of the universe does indeed resemble other scoops of the same size, fulfilling the long-sought aesthetic criterion for the cosmos. But, of course, on smaller scales, everything has clumped itself into distinctly non-homogeneous and non-isotropic distributions of matter.

Three centuries ago, Isaac Newton considered the question of how matter acquired structure. His creative mind easily embraced the concept of an isotropic and homogeneous universe, but promptly raised an issue that would not occur to most of us: How can you make any structure at all in the universe without having all the matter of the universe joining it to create one gigantic mass?

1: This map of the mottled cosmic background radiation was produced by NASA's Wilkinson Microwave Anisotropy Probe (WMAP). The slightly hotter regions of the sky are coded red in the image, and the slightly cooler regions blue. These deviations from an unchanging temperature everywhere betray variations in the density of matter during the earliest years of the universe. Superclusters of galaxies owe their origin to the slightly denser regions of this cosmic baby picture.

2: The Hubble Space Telescope's Ultra Deep Field, obtained in 2004, revealed the faintest cosmic objects ever recorded. Nearly every object in the image, no matter how small, is a galaxy, sitting anywhere from 3 to 10 billion light-years away from us. Because their light has traveled for billions of years before reaching the telescope, the galaxies appear not as they are today but as they once were, from their origins through the subsequent stages of their evolution.

3: This giant cluster of galaxies, called A2218 by astronomers, lies about 3 billion light-years from the Milky Way. Behind the galaxies in this cluster lie still more distant galaxies, whose light is bent and distorted primarily by the gravity from the dark matter and the most massive galaxies lurking within A2218. This bending produces the long, thin arcs of light visible in this image obtained by the Hubble Space Telescope.

4: Another giant cluster of galaxies, A1689, about 2 billion light-years away, also bends light from still more distant galaxies that happen to lie behind the cluster, producing short, bright arcs of light. By measuring the details of these arcs, revealed in images obtained by the Hubble Space Telescope, astronomers have determined that most of this cluster's mass resides not in the galaxies themselves, but in the form of dark matter.

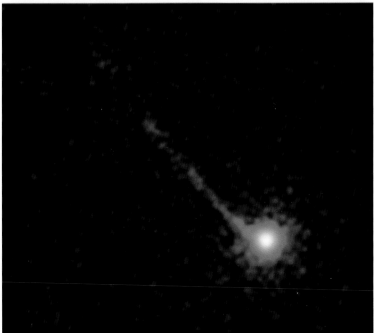

5: The quasar catalogued as PKS 1127-145 lies about 10 billion light-years from the Milky Way. In the top panel, a Hubble Space Telescope image in visible light, the quasar reveals itself as the bright object at the lower right. The actual quasar, which occupies only the innermost portion of this object, owes its enormous energy output to superheated matter falling into a titanic black hole. The bottom panel shows the same region of the sky in an X-ray image obtained by the Chandra Observatory. A jet of X-ray-emitting material more than a million light-years long spews forth from the quasar.

6: In this image of the Coma cluster of galaxies, nearly every faint object is in fact a galaxy made of more than 100 billion stars. Located about 325 million light-years from the Milky Way, this cluster spans a diameter of several million light-years and contains many thousand individual galaxies, orbiting one another in a kind of ballet choreographed by the forces of gravity.

7: The central region of the Virgo cluster of galaxies, a mere 60 million light-years from the Milky Way, shows dozens of galaxies of different types, including giant elliptical galaxies at the top left and top right of the image. Spiral galaxies appear throughout this image, taken with the Canada-France-Hawaii Telescope at the Mauna Kea Observatory. The Virgo cluster's immense gravitational force, and its proximity to the Milky Way, significantly affect the motion of the Milky Way through space. Indeed, the Milky Way and the Virgo cluster form part of an even larger system of galaxies called the Virgo *super*cluster.

8: This pair of interacting galaxies, named Arp 295 from their entry in Halton Arp's *Catalog of Peculiar Galaxies*, have drawn out long filaments of their own stars and gas, stretching across a quarter-million light-years. The two galaxies lie about 270,000 light-years from the Milky Way.

9: A giant spiral galaxy similar to our own dominates this photograph taken by the Very Large Telescope array in Chile. Our face-on view of this galaxy—about 100 million light-years from the Milky Way and named NGC 1232—allows us to observe the yellowish light from relatively old stars near the galaxy's center, as well as the massive hot, young, bluish stars that dominate the surrounding pinwheel of spiral arms. Astrophysicists also detect large numbers of interstellar dust grains within these arms. A smaller companion to NGC 1232, known as a barred spiral galaxy because its central regions have a barlike shape, appears to the left of the giant spiral.

10: This spiral galaxy, called NGC 3370 and about 100 million light-years away, closely resembles our own Milky Way in size, shape, and mass. This Hubble Space Telescope image reveals the complex spiral traced by the young, hot, highly luminous stars that outline the spiral arms. From rim to rim, the galaxy spans about 100,000 light-years.

11: In March 1994, astronomers discovered Supernova 1994D in the spiral galaxy NGC 4526, one of the thousands of galaxies in the Virgo cluster, about 60 million light-years from the Milky Way. In this image obtained by the Hubble Space Telescope, the supernova appears as the bright object at the lower left, below the belt of light-absorbing dust in the galaxy's central plane. Apart from enriching its environment with the chemical ingredients of life, Supernova 1994D is an example of the Type Ia supernovae used to discover the acceleration of the cosmic expansion.

12: When we look at this spiral galaxy, NGC 4631, about 25 million light-years away, our line of sight lies edge-on to the galaxy's disk, so we cannot see the galaxy's spiral-arm structure. Instead, dust that lies within the disk obscures much of the light from the galaxy's stars. The patch of red to the left of center marks a stellar nursery. Above NGC 4631 lies a smaller, elliptical galaxy, an orbiting companion to the giant spiral.

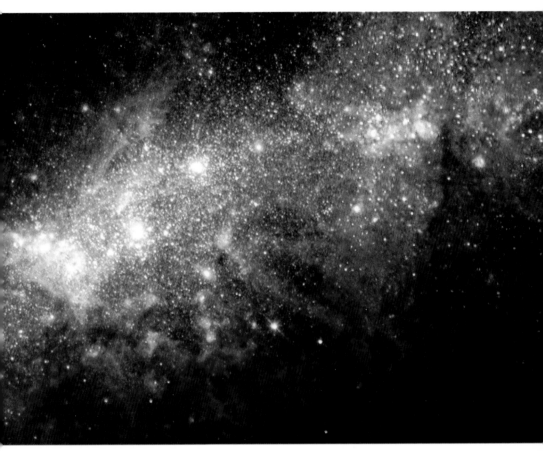

13: In this small irregular galaxy, called NGC 1569 and only 7 million light-years away, a burst of star formation began about 25 million years ago and can still be seen, accounting for most of the galaxy's light. Two large star clusters are visible in the left center of this Hubble Space Telescope image.

14: The Andromeda galaxy, the closest big galaxy to the Milky Way, lies about 2.4 million light-years from us and spans a region of the sky several times larger than the full moon. In this image, taken by amateur astronomer Robert Gender, one of the galaxy's two elliptical satellites appears below and to the left of its center, while a fainter one appears above and to the right of that center. All the other small bright objects in this image are individual stars within the Milky Way, sitting practically on our noses at less than 1/100 of the distance to the Andromeda galaxy.

15: Relatively close to the Milky Way, at about the same distance as the Andromeda galaxy (2.4 million light-years), lies the smaller spiral galaxy M33, whose largest star-forming region appears in this Hubble Space Telescope image. The most massive stars to form in this region have already exploded as supernovae, enriching their environment with heavy elements, while other massive stars are producing intense ultraviolet radiation that blasts electrons from the atoms surrounding them.

16: The Milky Way has two large irregular satellite galaxies, called the Large and Small Magellanic Clouds. This image of the Large Magellanic Cloud shows a large bar of stars at the left, with many additional stars and star-forming regions to the right. The bright Tarantula Nebula, named for its shape and seen at the upper center of the photograph, is the largest star-forming region in this galaxy.

17: This star-forming region, called the Papillon nebula for its resemblance to a butterfly, belongs to the Large Magellanic Cloud, the Milky Way's largest satellite galaxy. Young stars illuminate the nebula from inside and excite hydrogen atoms so they emit a characteristic shade of red, captured in this image by the Hubble Space Telescope.

18: A survey of the entire sky in infrared radiation reveals that we live inside the flattened disk of a spiral galaxy, which extends in this image to the left and right of the Milky Way's central region. Dust particles absorb some of the light from this region, just as they do in faraway spiral galaxies. Below the plane of our galaxy we can see the Milky Way's two irregular satellite galaxies, the Large and Small Magellanic Clouds.

19: When we look toward the center of our Milky Way galaxy, about 30,000 light-years from the solar system, enormous dust-rich clouds block our view in visible light. Infrared light does a better job of penetrating the dust, so this infrared image obtained from the Two Micron All Sky Survey project reveals radiation that arises close to the galactic center, the particularly bright region in this image, where a supermassive black hole may be steadily swallowing matter.

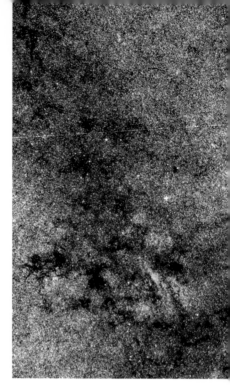

20: The Crab nebula lives about 7,000 light-years from the solar system, and was produced by an exploding star whose light reached the Earth on July 4, 1054. In this image taken by the Canada-France-Hawaii Telescope at the Mauna Kea Observatory, the reddish filaments consist primarily of hydrogen gas, expanding away from the region of the explosion at the center. The whitish glow arises from electrons moving at nearly the speed of light through intense magnetic fields. Supernova remnants such as this one add their evolved material to interstellar clouds of gas and dust. These clouds give birth to new stars that contain more "heavy" elements such as carbon, nitrogen, oxygen, and iron than older stars do.

Newton argued that since we observe no such mass the universe must be infinite. In 1692, writing to Richard Bentley, the master of Trinity College at Cambridge University, Newton proposed that

> if all the matter in the universe were evenly scattered throughout all the heavens, and every particle had an innate gravity toward all the rest, and the whole space throughout which this matter was scattered was but finite, the matter on the outside of the space would, by its gravity, tend toward all the matter on the inside, and by consequence, fall down into the middle of the whole space and there compose one great spherical mass. But if the matter was evenly disposed throughout an infinite space, it could never convene into one mass; but some of it would convene into one mass and some into another, so as to make an infinite number of great masses, scattered at great distances from one to another throughout all that infinite space.

Newton presumed that his infinite universe must be static, neither expanding nor contracting. Within this universe, objects were "convened" by gravitational forces—the attraction that every object with mass exerts on all other objects. His conclusion about gravity's central role in creating structure remains valid today, even though cosmologists face a task more daunting than Newton's. Far from enjoying the benefits of a static universe, we must allow for the fact that the universe has been expanding ever since the big bang, naturally opposing any tendency for matter to clump together by gravity. The problem of overcoming the cosmic expansion's anti-convening tendency becomes more serious when we consider that the cosmos expanded most rapidly soon after the big bang, the era when structures first began to form. At first glance, we could no more rely on gravity to form massive objects out of diffuse gas than we could use a shovel to move fleas across a barnyard. Yet somehow gravity has done the trick.

During the early days of the universe, the cosmos expanded so rapidly that if the universe had been strictly homogenous and isotropic on all size scales, gravity would have had no chance of victory. Today these would be no galaxies, stars, planets, or people, only a scattered distribution of atoms everywhere in space—a dull and boring cosmos, devoid of admirers and objects of admiration. But ours is a fun and exciting universe only because *inho*mogeneities and *an*isotropies appeared during those earliest cosmic moments, which served as a kind of cosmic soup-starter for all concentrations of matter and energy that would later emerge. Without this head start, the rapidly expanding universe would have prevented gravity from ever gathering matter to build the familiar structures we take for granted in the universe today.

What made these deviations, the inhomogeneities and anisotropies that provide the seeds for all the structure in the cosmos? The answer arrives from the realm of quantum mechanics, undreamt of by Isaac Newton but unavoidable if we hope to understand where we came from. Quantum mechanics tells us that on the smallest scales of size, no distribution of matter can remain homogeneous and isotropic. Instead, random fluctuations in the distribution of matter will appear, disappear, and reappear in different amount, as matter becomes a quivering mass of vanishing and reborn particles. At any particular time, some regions of space will have slightly more particles, and therefore a slightly greater density, than other regions. From this counterintuitive, airy-fairy fantasy, we derive everything that exists. The slightly denser regions had the chance to attract slightly more particles by gravity, and with time the cosmos grew these denser regions into structures.

In tracing the growth of structure from times soon after the big bang, we can gain some insight from two key epochs we have already met, the "era of inflation," when the universe expanded at an astounding rate, and the "time of decoupling," about

380,000 years after the big bang, when the cosmic background radiation ceased to interact with matter.

The inflationary era lasted from about 10^{-37} second to 10^{-33} second after the big bang. During that relatively brief stretch of time, the fabric of space and time expanded faster than light, growing in a billionth of a trillionth of a trillionth of a second from one hundred billion billion times smaller than the size of a proton to about 4 inches. Yes, the observable universe once fit within a grapefruit. But what caused the cosmic inflation? Cosmologists have named the culprit: a "phase transition" that left behind a specific and observable signature in the cosmic background radiation.

Phase transitions are hardly unique to cosmology; they often occur in the privacy of your home. We freeze water to make ice cubes, and boil water to produce steam. Sugary water grows sugar crystals on a string dangling within the liquid. And wet, gooey batter turns into cake when baked. There's a pattern here. In every case, things look very different on the two sides of a phase transition. The inflationary model of the universe asserts that when the universe was young, the prevailing energy field went through a phase transition, one of several that would have occurred during these early times. This particular episode not only catapulted the early, rapid expansion but also imbued the cosmos with a specific fluctuating pattern of high- and low-density regions. These fluctuations then froze into the expanding fabric of space, creating a kind of blueprint for where galaxies would ultimately form. Thus in the spirit of Pooh-Bah, the character in Gilbert and Sullivan's *Mikado* who proudly traced his ancestry back to a "primordial atomic globule," we can assign our origins, and the beginnings of all structure, to the fluctuations on a subnuclear scale that arose during the inflationary era.

What facts can we cite to support this bold assertion? Since astrophysicists have no way to see back to the universe's first 0.0000000000000000000000000000000000001 of a second, they do the next best thing, and use scientific logic to connect this early epoch to times they can observe. If the inflationary theory is correct, the initial fluctuations produced during that era, the inevitable result of quantum mechanics—which tells us that small variations from place to place will always arise within an otherwise homogeneous and isotropic fluid—would have had the opportunity to become regions of high and low concentrations of matter and energy. We can hope to find evidence for these variations from place to place in the cosmic background radiation, which serves as a proscenium that separates the current epoch from, and also connects it with, the first moments of the neonate universe.

As we have already seen, the cosmic background radiation consists of the photons generated during the first minutes after the big bang. Early in the universe's history, these photons interacted with matter, slamming into any atoms that happened to form so energetically that no atoms could exist for long. But the ongoing expansion of the universe in effect robbed the photons of energy, so that eventually, at the time of decoupling, none of the photons had energies sufficient to prevent electrons from orbiting around protons and helium nuclei. Since that time, 380,000 years after the big bang, atoms have persisted—unless some local disturbance, such as the radiation from a nearby star, disrupts them— while the photons, each with an ever-diminishing amount of energy, continue to roam the universe, collectively forming the cosmic background radiation or CBR.

The CBR thus carries the imprint of history, a snapshot of what the universe was like at the time of decoupling. Astrophysicists have learned how to examine this snapshot with ever-increasing accuracy. First, the simple fact that the CBR exists, that their basic understanding of the history of the universe is correct. And then, after years of improving their abilities to measure the cosmic background radiation, their sophisticated balloon-borne and satellite instruments gave them a map of the CBR's tiny deviations from homogeneity. This map provides the record of the once minuscule fluctuations whose size increased as the universe expanded during the few hundred thousand years after the era of inflation, and which then grew, during the next billion years or so, into the large-scale distribution of matter in the cosmos.

Remarkable though it may seem, the CBR provides us with the means for mapping the imprint of the long-vanished early universe, and for locating—14 billion light-years away in all directions—the regions of slightly greater density that would become galaxy clusters and superclusters. Regions with greater-than-average density left behind slightly more photons than regions with lower densities. As the cosmos became transparent, thanks to the loss of energy that left the photons unable to interact with the newly formed atoms, each photon embarked on a journey that would carry it far from its point of origin. Photons from our vicinity have traveled 14 billion light-years in all directions, providing part of the CBR that far-distant civilizations at the end of the visible universe may even now be examining, and "their" photons, having reached our instruments, tell us about what things were like long ago and far away, in the times when structures had barely begun to form.

Through more than a quarter of a century following the first detection of the cosmic background radiation in 1965, astrophysicists searched for anisotropies in the CBR. From a theoretical viewpoint, they desperately needed to find them, because without the existence of CBR anisotropies at the level of a few parts in a hundred thousand, their basic model of how structure appeared would lose all claim to validity. Without the seeds of matter they betray, we would have no explanation for why we exist. As happy fate would have it, the anisotropies appeared precisely on schedule. Just as soon as cosmologists created instruments capable of detecting anisotropies at the appropriate level, they found them, first with the COBE satellite in 1992, and later with far more precise instruments mounted on balloons and on the WMAP satellite described in Chapter 3. The teeny fluctuations from place to place in the amounts of microwave photons that form the CBR, now delineated with impressive precision by WMAP, embody the record of cosmic fluctuations at a time 380,000 years after the big bang. The typical fluctuation sits only a few hundred thousandths of a degree above or below the average temperature of the cosmic background radiation, so detecting them is like finding faint spots of oil on a mile-wide pond that make the water plus oil a tiny bit denser than average. Small though these anisotropies were, they sufficed to get things started.

In the WMAP map of the cosmic background radiation, the larger hot spots tell us where gravity would overcome the expanding universe's dissipative tendencies and gather together enough matter to manufacture superclusters. These regions today have grown to contain about 1,000 galaxies, each with 100 billion stars. If we add the dark matter in such a supercluster, its total mass reaches the equivalent of 10^{16} Suns. Conversely, the larger cool spots, with no head start against the expanding universe, evolved to become nearly devoid of massive structures. Astrophysicists just call these regions "voids," a term that gains meaning from being surrounded by something that is not a void. So the giant sheets and filaments of galaxies that we can trace on the sky not only form clusters at their intersections but also trace walls and other geometric forms that give shape to the empty regions of the cosmos.

Of course, the galaxies themselves did not simply appear, fully formed, from concentrations of matter a tiny bit denser than average. From 380,000 years after the big bang until about 200 million years later, matter continued to gather itself together, but nothing shone in the universe, whose first stars were yet to be born. During this cosmic dark age, the universe contained only what it had made during its first few minutes—hydrogen and helium, with traces of lithium. With no elements heavier than these—no carbon, nitrogen, oxygen, sodium, calcium, or heavier elements—the cosmos contained none of the now common atoms or molecules that can absorb light as a star begins to shine. Today, in the presence of these atoms and molecules, the light from a newly formed star will exert pressure upon them that pushes away massive quantities of gas that would otherwise fall into the star. This expulsion limits the maximum mass of newborn stars to less than one hundred times the Sun's mass. But when the first stars formed, in the absence of atoms and molecules that would absorb starlight, infalling gas consisted almost entirely of hydrogen and helium, providing only token resistance to stars' output. This allowed stars to form with much larger masses, up to many hundred, perhaps even a few thousand, times the mass of the Sun.

High-mass stars live life in the fast lane, and the most massive live the most rapidly of all. They convert their matter into energy at astonishing rates, as they manufacture heavy elements and die explosive youthful deaths. Their life expectancies amount to no more than a few million years, less than a thousandth of the Sun's. We expect to find none of the most massive stars from that era alive today, because the early ones burnt themselves out long ago, and today, with heavier elements common throughout the universe, the highest-mass stars of old cannot form at all. Indeed, none of the high-mass giants has ever been observed. But we assign them the responsibility for having first introduced into the universe almost all of the familiar elements we now take for granted, including carbon, oxygen, nitrogen, silicon, and iron. Call it enrichment. Call it pollution. But the seeds of life began with the long-vanished first generation of high-mass stars.

During the first few billion years after the time of decoupling, gravitationally induced collapse proceeded with abandon, as grav-

ity drew matter together on nearly all scales. One of the natural
results of gravity at work was the formation of supermassive
black holes, each with a mass millions or billions of times the
mass of the Sun. Black holes with that amount of mass are about
the size of Neptune's orbit and wreak havoc on their nascent environment. Gas clouds drawn toward these black holes want to gain
speed, but they can't, because there's too much stuff in the way.
Instead, they slam into and rub against whatever came in just
before them, descending toward their master in a swirling mael-
strom. Just before these clouds disappear forever, collisions within
their superheated matter radiate titanic quantities of energy, billions of times the Sun's luminosity, all within the volume of a
solar system. Monstrous jets of matter and radiation spew forth,
extending hundreds of thousands of light-years above and below
the swirling gas, as the energy punches through and escapes the
funnel in all ways it can. As one cloud falls, and another orbits-in-
waiting, the luminosity of the system fluctuates, getting brighter
and dimmer over a matter of hours, days, or weeks. If the jets
happen to be aimed straight at you, the system will look even
more luminous, and more variable in its output, than those cases
in which the jets point to the side. Viewed from any appreciable
distance, all of these black hole–plus–infalling-matter combinations will appear amazingly small and luminous in comparison to
the galaxies we see today. What the universe has created—the
objects whose birth we have just witnessed in words—are quasars.

Quasars were discovered during the early 1960s, as astronomers
began to use telescopes equipped with detectors sensitive to invisible domains of radiation, such as radio waves and X-rays. Their
galaxy portraits could therefore include information about the
galaxies' appearance in those other bands of the electromagnetic
spectrum. Combine this with further improvements in photographic emulsions, and a new zoo of galaxy species emerged from
the depths of space. Most remarkable among them were objects

that, in photographs, look like simple stars, but—quite unlike stars—produce extraordinary amounts of radio waves. The working description for those objects was "quasistellar radio source"— a term quickly shortened to quasar. Even more remarkable than the radio emission from these objects were their distances: as a class, they turned out to be the most distant objects known in the universe. For quasars to be that small and still visible at immense distances meant that they had to be an entirely new kind of object. How small? No bigger than a solar system. How luminous? Even the dim ones outshine your average galaxy in the universe.

By the early 1970s, astrophysicists had converged on supermassive black holes as the quasar engine, gravitationally devouring everything in its grasp. The black hole model can account for how small and bright quasars are, but says nothing of the black hole's source of food. Not until the 1980s would astrophysicists begin to understand the quasar's environment, because the tremendous luminosity of a quasar's central regions prevents any sight of its much fainter surroundings. Eventually, however, with new techniques to mask the light from the center, astrophysicists could detect fuzz surrounding some of the dimmer quasars. As detection tactics and technologies improved further, every quasar revealed fuzz; some even revealed a spiral structure. Quasars, it turned out, are not a new kind of object but rather a new kind of galactic nucleus.

In April 1991, the National Aeronautics and Space Administration (NASA) launched one of the most expensive astronomical instruments ever built: the Hubble Space Telescope. The size of a Greyhound bus, directed by commands sent from Earth, the Hubble Telescope could profit from orbiting outside our ever-blurring atmosphere. Once astronauts had installed lenses to correct for mistakes in the way its primary mirror had been made, the telescope could peer into previously uncharted regions of ordinary galaxies, including their centers. Upon gazing into those centers, it found the stars moving inexcusably fast, given the gravity inferred from the visible light of other stars in the vicinity. Hmmm, strong gravity, small area . . . must be a black hole. Galaxy after galaxy—dozens of them—had suspiciously speedy stars in their cores. Indeed, whenever the Hubble Space Telescope had a clear view of a galaxy's center, they were there.

It now seems likely that every giant galaxy harbors a super-massive black hole, which could have served as a gravitational seed around which the other matter collected or may have been manufactured later by matter streaming down from outer regions of the galaxy. But not all galaxies were quasars in their youth.

The growing roster of ordinary galaxies known to have a black hole at their center began to raise eyebrows among investigators: A supermassive black hole that was not a quasar? A quasar that's surrounded by a galaxy? One can't help but think of a new picture of how things work. In this picture, some galaxies begin their lives as quasars. To be a quasar, which is really just the blazing vis-ible core of an otherwise run-of-the-mill galaxy, the system has to have not only a massive, hungry black hole but also an ample sup-ply of infalling gas. Once the supermassive black hole has gulped down all the available food, leaving uneaten stars and gas in dis-tant, safe orbits, the quasar simply shuts off. You've then got a docile galaxy with a dormant black hole snoozing at its center.

Astronomers have found other new types of objects, classified as intermediate between quasars and normal galaxies, whose properties also depend on the bad behavior of supermassive black holes. Sometimes the streams of material falling into a galaxy's central black hole flow slowly and steadily. At other times episodically. Such systems populate the menagerie of galaxies whose nuclei are active but not ferocious. Over the years, names for the various types accumulated: LINERs (low-ionization nuclear emission-line regions), Seyfert galaxies, N galaxies, blazars. All of these objects are generically called AGNs, the astrophysicist's abbreviation for galaxies with "active" nuclei. Unlike quasars, which appear only at immense distances, AGNs appear both at large distances and relatively nearby. This suggests that AGNs fill in the range of galaxies that misbehave. Quasars long ago consumed all their food, so we see them only when we look far back in time by observing far out in space. AGNs, in contrast, had more modest appetites, so some of them still have food to eat even after billions of years.

Classifying AGNs solely on the basis of their visual appearance alone would provide an incomplete story, so astrophysicists classified AGNs by their spectra and by the full range of their electromagnetic emissions. During the mid- to late 1990s, investigators improved their black hole model, and found that they could characterize nearly all the beasts in the AGN zoo by measuring only a few parameters: the mass of the object's black hole, the rate at which it's being fed, and our angle of view on the accretion disk and its jets. If, for example, we happen to look "right down the barrel," along exactly the same direction as that of a jet emerging from the vicinity of a supermassive black hole, we see a much brighter object than if we happen to have a side view from a much different angle. Variations in these three parameters can account for nearly all the impressive diversity that astrophysicists observe, giving them a welcome de-speciation of galaxy types and a deeper understanding of the formation and evolution of galaxies. The fact that so much can be accounted for—differences in shape, size, luminosity, and color—by so few variables represents an unheralded triumph of late twentieth-century astrophysics.

Because it took a lot of investigators and a lot of years and a lot of telescope time, it's not the sort of thing that gets announced on the evening news—but it's a triumph nonetheless.

Let us not conclude, however, that supermassive black holes can explain everything. Even though they have millions or billions of times the Sun's mass, they contribute almost nothing in comparison with the masses of the galaxies in which they are embedded—typically far less than 1 percent of a large galaxy's total mass. When we seek to account for the existence of dark matter, or of other unseen sources of gravity in the universe, these black holes are insignificant and may be ignored. But when we calculate how much energy they wield—that is, when we compute the energy that they released as part of their formation—we find that black holes dominate the energetics of galaxy formation. All the energy of all the orbits of all the stars and gas clouds that ultimately compose a galaxy pales when compared with what made the black hole. Without supermassive black holes lurking below, galaxies as we know them might have never formed. The once luminous but now invisible black hole that lies at the center of each giant galaxy provides a hidden link, the physical explanation for the agglomeration of matter into a complex system of billions of stars in orbit around a common center.

The broader explanation for the formation of galaxies invokes not only the gravity produced by supermassive black holes but also gravity in more conventional astronomical settings. What made the billions of stars in a galaxy? Gravity did this too, producing up to hundreds of thousands of stars in a single cloud. Most of a galaxy's stars were born within relatively loose "associations." The more compact regions of starbirth remain identifiable "star clusters," within which member stars orbit the cluster's center, tracing their paths through space in a cosmic ballet choreographed by the forces of gravity from all the other stars within the cluster, even as the clusters themselves move on enormous trajectories around the galactic center, safe from the destructive power of the central black hole.

Within a cluster, stars move at a broad range of speeds, some so rapidly that they risk escape from the system altogether. This indeed occasionally occurs, as fast stars evaporate from the grip of a cluster's gravity to roam freely through the galaxy. These free-ranging stars, along with the "globular star clusters" that contain hundreds of thousands of stars each, add to the stars that form the spherical haloes of galaxies. Initially luminous, but today devoid of their brightest, short-lived stars, galaxy haloes are the oldest visible objects in the universe, with birth certificates traceable to the formation of galaxies themselves.

Last to collapse, and thus the last to turn into stars, we encounter the gas and dust that finds itself pulled and pinned into the galactic plane. In elliptical galaxies, no such plane exists, and all of their gas has already turned into stars. Spiral galaxies, however, have highly flattened distributions of matter, characterized by a central plane within which the youngest, brightest stars form in spiral patterns, testimony to great vibrating waves of alternating dense and rarefied gas that orbit the galactic center. Like hot marshmallows that stick together upon contact, all of the gas in a spiral galaxy that did not swiftly participate in making star clusters has fallen toward the galactic plane, stuck to itself, and created a disk of matter that slowly manufactures stars. For past billions of years, and for billions of years to come, stars will continue to form in spiral galaxies, with each generation more enriched in heavy elements than the next. These heavy elements (by which astrophysicists mean all elements heavier than helium) have been cast forth into interstellar space by outflows from aging stars or as the explosive remains of high-mass stars, a species of supernova. Their existence renders the galaxy—and thus the universe—ever more friendly to the chemistry of life as we know it.

We have outlined the birth of a classical spiral galaxy, in an evolutionary sequence that has played out tens of billions of times, yielding galaxies in a host of different arrangements: In clusters of galaxies. In long strings and filaments of galaxies. And in sheets of galaxies.

Because we look back in time as we look outward into space, we possess the ability to examine galaxies not only as they are now but also as they appeared billions of years ago, simply by looking up. The problem with turning this concept into observational reality resides in the fact that galaxies billions of light-years away appear to us as extremely small and dim objects, so even our best telescopes can barely resolve their outlines. Nevertheless, astrophysicists have made great progress in this effort during the past few years. The breakthrough came in 1995, when Robert Williams, then the director of the Space Telescope Science Institute at Johns Hopkins University, arranged for the Hubble Telescope to point toward a single direction in space, near the Big Dipper, for ten days' worth of observation. Williams deserves the credit because the telescope's Time Allocation Committee, which selects the observing proposals most worthy of actual telescope time, judged it unworthy of support. After all, the region to be studied was deliberately chosen for having nothing interesting to look at, and thus to represent a dull and boring patch of sky. As a result, no ongoing projects could benefit directly from such a large commitment of the telescope's highly oversubscribed observing time. Happily, Williams, as the director of the Space Telescope Science Institute, had the right to assign a few percent of the total—his "director's discretionary time"—and invested his clout on what became known as the Hubble Deep Field, one of the most famous astronomical photographs ever taken.

The ten-day exposure, coincidentally made during the government shutdown of 1995, produced by far the most researched image in the history of astronomy. Studded with galaxies and galaxylike objects, the deep field offers us a cosmic palimpsest, in which objects at different distances from the Milky Way have written their momentary signatures of light at different times. We see objects in the deep field as they were, say, 1.3 billion, 3.6 billion, 5.7 billion, or 8.2 billion years ago, with each object's epoch determined by its distance from us. Hundreds of astronomers have seized upon the wealth of data contained in this single image to derive new information about how galaxies have evolved with time, and about how galaxies looked soon after they formed. In 1998, the telescope secured a companion image, the Hubble Deep Field South, by devoting ten days of observation to another patch of sky in the direction opposite to that of the first deep field, in the celestial southern hemisphere. Comparison of the two images allowed astronomers to assure themselves that the results from the first deep field did not represent an anomaly (for example, if the two images had been identical in every detail, or statistically unlike each other in every way, one might have concluded that the devil was at work), and to refine their conclusions about how different types of galaxies form. After a successful servicing mission, in which the Hubble Telescope was outfitted with even better (more sensitive) detectors, the Space Telescope Science Institute just couldn't resist and, in 2004, authorized the Hubble Ultra Deep Field, laying bare the ever more distant cosmos.

Unfortunately, the earliest stages of galaxy formation, which would be revealed to us by objects at the greatest distances, confound even the Hubble Telescope's best efforts, not least because the cosmic expansion has shifted most of their radiation into the infrared region of the spectrum, not accessible to the telescope's instruments. For these most distant galaxies, astronomers await the design, construction, launch, and successful operation of the Hubble's successor, the James Webb Space Telescope (JWST), named after the head of NASA during the Apollo era. (Cynics say that this name, rather than one that honors a famous scientist, was chosen to assure that the telescope project will not be canceled, since this would involve deleting an important official's legacy.)

The JWST will have a mirror larger than Hubble's, designed to unfurl itself like an intricate mechanical flower, opening in space to provide a reflective surface much larger than any that can fit inside one of our rockets. The new space telescope will also possess a suite of instruments far superior to those of the Hubble Telescope, which were originally designed during the 1960s, built during the 1970s, launched in 1991, and—even though significantly upgraded during the 1990s—still lack such fundamental abilities as the capacity to detect infrared radiation. Some of this ability now exists in the Spitzer InfraRed Telescope Facility (SIRTF), launched in 2003, which orbits the Sun much farther from Earth than the Hubble does, thereby avoiding interference from the copious amounts of infrared radiation produced by our planet. To achieve this goal, JWST will likewise have an orbit much farther from Earth than the Hubble Telescope does, and will therefore be forever inaccessible to servicing missions as they are currently conceived—NASA had better get this one right the first time. If the new telescope goes into operation in 2011, as currently planned, it should then provide spectacular new views of the cosmos, including images of galaxies more than 10 billion light-years away, seen much closer to their time of origin than any revealed by the Hubble Deep Fields. Working in tandem with the new space telescope, as they have with the old, large ground-based instruments will study in detail the wealth of objects to be revealed by our next great step in space-borne instrumentation.

Rich in possibility though the future may be, we should not neglect the astrophysicists' impressive accomplishments during the past three decades, which spring from their abilities to create new instruments to observe the universe. Carl Sagan liked to say that you had to be made from wood not to stand in awe of what the cosmos has done. Thanks to our improved observations, we now know more than Sagan did about the amazing sequence of events that led to our existence: the quantum fluctuations in the distribution of matter and energy on a scale smaller than the size of a proton that spawned superclusters of galaxies, thirty million light-years across. From chaos to cosmos, this cause-and-effect relationship crosses more than thirty-eight powers of ten in size and forty-two powers of ten in time. Like the microscopic strands of DNA that predetermine the identity of a macroscopic species and the unique properties of its members, the modern look and feel of the cosmos was writ in the fabric of its earliest moments, and carried relentlessly through time and space. We feel it when we look up. We feel it when we look down. We feel it when we look within.

Part III

The Origin
of Stars

CHAPTER 9

Dust to Dust

f you look at the clear night sky far from city lights, you can immediately locate a cloudy band of pale light, broken in places by dark splotches, that runs from horizon to horizon. Long known as the (lower-case) "milky way" in the sky, this milk-white haze combines the light from a staggering number of stars and gaseous nebulae. Those who observe the milky way with binoculars or a backyard telescope will see the dark and boring areas resolve themselves into, well, dark and boring areas—but the bright areas will turn from a diffuse glow into countless stars and nebulae.

In his small book *Sidereus Nuncius (The Starry Messenger)*, published in Venice in 1610, Galileo Galilei provided the first account of the heavens as seen through a telescope, including a description of the milky way's patches of light. Referring to his instrument as a spyglass, since the name telescope ("far-seer" in Greek) had yet to be coined, Galileo could barely contain himself:

The milky way itself, which, with the aid of the spyglass, may be observed so well that all the disputes that for so many generations have vexed philosophers are destroyed by visible certainty, and we are liberated from wordy arguments. For the Galaxy is nothing else than a congeries of innumerable stars distributed in clusters. To whatever region of it you direct your spyglass, an immense number of stars immediately offer themselves to view, of which very many appear rather large and very conspicuous but the multitude of small ones is truly unfathomable.[*]

Surely Galileo's "immense number of stars," which delineate the most densely packed regions of our Milky Way galaxy, must locate the real astronomical action. Why, then, should anybody be interested in the intervening dark areas with no visible stars? Based on their visual appearance, the dark areas are probably cosmic holes, openings to the infinite and empty spaces beyond.

Three centuries would pass before anyone figured out that the dark patches in the milky way, far from being holes, actually consist of dense clouds of gas and dust that obscure more distant star fields and hold stellar nurseries deep within themselves. Following earlier suggestions by the American astronomer George Cary Comstock, who wondered why faraway stars are much dimmer than their distances alone would indicate, the Dutch astronomer Jacobus Cornelius Kapteyn in 1909 identified the culprit. In two research papers, both titled "On the Absorption of Light in Space,"[†] Kapteyn presented evidence that the dark clouds—his newfound "interstellar medium"—not only block the light from stars but also do so unevenly across the rainbow of colors in a star's spectrum: they absorb and scatter, and therefore attenuate, light at

[*] Galileo Galilei, *Siderius Nuncius*, trans. Albert van Helden (Chicago: University of Chicago Press, 1989), p. 62.

[†] J. C. Kapteyn, *Astrophysical Journal* 29, 46, 1909; 30, 284, 1909.

the violet end of the visible spectrum more effectively than they act on red light. This selective absorption preferentially removes more violet than red light, making faraway stars appear redder than nearby ones. The amount of this interstellar reddening of starlight increases in proportion to the total amount of material that the light encounters on its journey to us.

Ordinary hydrogen and helium, the principal constituents of cosmic gas clouds, don't redden light. But molecules made of many atoms do so—especially those that contain the elements carbon and silicon. When interstellar particles grow too large to be called molecules, with hundreds of thousand or millions of individual atoms in each of them, we call them dust. Most of us know dust of the household variety, although few of us care to learn that, in a closed home, dust consists mostly of dead, sloughed-off human skin cells (plus pet dander, if you have one or more live-in mammals). As far as we know, cosmic dust contains nobody's epidermis. However, interstellar dust does include a remarkable ensemble of complex molecules, which emit photons primarily in the infrared and microwave regions of the spectrum. Astrophysicists lacked good microwave telescopes until the 1960s, and effective infrared telescopes until the 1970s. Once they had created these observational instruments, they could investigate the true chemical richness of the stuff that lies between the stars. During the decades that followed these technological advances, a fascinating, intricate picture of star birth emerged.

Not all gas clouds will form stars at all times. More often than not, a cloud finds itself confused about what to do next. Actually, astrophysicists are the confused ones here. We know that an interstellar cloud "wants" to collapse under its own gravity to make one or more stars. But the cloud's rotation, as well as the effects of turbulent gas motions within the cloud, oppose that result. So, too, does the gas pressure that you learned about in high school chemistry class. Magnetic fields can also fight collapse. They penetrate the cloud and constrain the motions of any free-roaming charged particles contained therein, resisting compression and thus impeding the ways in which the cloud can respond to its own gravity. The scary part of this thought-exercise comes from the realization that if no one knew in advance that stars exist, front-line research would offer plenty of convincing reasons why stars could never form.

Like the several hundred billion stars in our Milky Way galaxy, named after the band of light that the galaxy's most densely populated regions paint across our skies, giant clouds of gas orbit our galaxy's center. The stars amount to tiny specks, only a few light-seconds across, that float in a vast ocean of nearly empty space, occasionally passing close by one another like ships in the night. Gas clouds, on the other hand, are huge. Typically spanning hundreds of light-years, they each contain as much mass as a million Suns. As these giant clouds lumber through the galaxy, they often collide with one another, entangling their gas- and dust-laden innards. Sometimes, depending on their relative speeds and their angles of impact, the clouds stick together; at other times, adding injury to the insult of collision, they rip each other apart.

If a cloud cools to a sufficiently low temperature (less than about 100 degrees above absolute zero), its constituent atoms will stick together when they collide, rather than careening off one another as they do at higher temperatures. This chemical transition has consequences for everybody. The growing particles—now containing tens of atoms each—begin to scatter visible light to and fro, strongly attenuating the light of the stars behind the cloud. By the time that the particles become full-grown dust grains, they each contain billions of atoms. Aging stars manufacture similar dust grains and blow them gently into interstellar space during their "red-giant" phases. Unlike smaller particles, dust grains with billions of atoms no longer scatter the visible light photons from the stars behind them; instead, they absorb those photons and then reradiate their energy as infrared, which can easily escape from the cloud. As this occurs, the pressure from the photons, transmitted to the molecules that absorb it, pushes the cloud in the direction opposite to the direction of the light source. The cloud has now coupled itself to starlight.

Star birth occurs when the forces that make a cloud progressively denser eventually lead to its gravitationally induced collapse, during which each part of the cloud pulls all the other parts much closer. Since hot gas resists compression and collapse more effectively than cool gas does, we face an odd situation. We must cool the cloud before it can ever heat itself by producing a star. In other words, the creation of a star that possesses a 10-million-degree core, sufficiently hot for thermonuclear fusion to begin, requires that the cloud must first achieve its coldest possible internal conditions. Only at extremely cold temperatures, a few dozen degrees above absolute zero, can the cloud collapse and allow star formation to begin in earnest.

What happens within a cloud to turn its collapse into newborn stars? Astrophysicists can only gesticulate. Much as they would like to track the internal dynamics of a large, massive interstellar cloud, the creation of a computer model that includes the laws of physics, all the internal and external influences on the cloud, and all the relevant chemical reactions that can occur within it still lies beyond our abilities. A further challenge resides in the humbling fact that the original cloud has a size billions of times larger than that of the star we are trying to create—which in turn has a density 100 sextillion times the average density within in the cloud. In these situations, what matters most on one scale of sizes may not be the right thing to worry about on another.

Nevertheless, relying on what we see throughout the cosmos, we can safely assert that within the deepest, darkest, densest regions of an interstellar cloud, where temperatures fall to about 10 degrees above absolute zero, gravity does cause pockets of gas to collapse, easily overcoming the resistance offered by magnetic fields and other impediments. The contraction converts the cloud pockets' gravitational energy into heat. The temperature within each of these regions—soon to become the core of a newborn star—rises rapidly during the collapse, breaking apart all the dust grains in the immediate vicinity as they collide. Eventually, the temperature in the central region of the collapsing gas pocket reaches the crucial value of 10 million degrees on the absolute scale.

At this magic temperature, some of the protons (which are simply naked hydrogen atoms, shorn of the electron that orbits them) move fast enough to overcome their mutual repulsion. Their high speeds allow the protons to approach one another closely enough for the "strong nuclear force" to make them bond. This force, which operates only at extremely short distances, binds together the protons and neutrons in all nuclei. The thermonuclear fusion of protons—"thermo" because it occurs at high temperatures, and "nuclear fusion" because it fuses particles into a single nucleus—creates helium nuclei, each of which has a mass slightly less than the sum of the particles from which it fused. The mass that disappears during this fusion turns into energy, in a balance described by Einstein's famous equation. The energy embodied in mass (always in an amount equal to the mass times the square of the speed of light) can be converted into other forms of energy, such as additional kinetic energy (energy of motion) of the fast-moving particles that emerge from nuclear fusion reactions.

As the new energy produced by nuclear fusion diffuses outward, the gas heats and glows. Then, at the star's surface, the energy formerly locked in individual nuclei escapes into space in the form of photons, generated by the gas as the energy released through fusion heats it to thousands of degrees. Even though this region of hot gas still resides within the cosmic womb of a giant interstellar cloud, we may nonetheless announce to the Milky Way that . . . a star is born.

Astronomers know that stars range in mass from a mere one tenth of the Sun's to nearly one hundred times our star's mass. For reasons not well understood, a typical giant gas cloud can develop a multitude of cold pockets that all tend to collapse at about the same time to give birth to stars—some puny and others giants. But the odds favor the puny: for every high-mass star, a thousand low-mass stars are born. The fact that no more than a few percent of all the gas in the original cloud participates in star birth presents a classic challenge in explaining star formation: What makes the star-forming tail wag the largely unchanged dog of an interstellar gas cloud? The answer probably lies in the radiation produced by newborn stars, which tends to inhibit further star formation.

We can easily explain the lower bound on the masses of newborn stars. Pockets of collapsing gas with masses less than about one tenth of the Sun's have too little gravitational energy to raise their core temperatures to the 10 million degrees required for the nuclear fusion of hydrogen. In that case, no nuclear-fusing star will be born; instead, we obtain a failed, would-be star—an object that astronomers call a "brown dwarf." With no energy source of its own, a brown dwarf fades steadily, shining from the modest heat generated during the original collapse. The gaseous outer layers of a brown dwarf are so cool that many of the large molecules normally destroyed in the atmospheres of hotter stars remain alive and well within them. Their feeble luminosities make brown dwarfs immensely difficult to detect, so to find them, astrophysicists must employ complex methods similar to those they occasionally use to detect planets: searching for the faint infrared glow from these objects. Only in recent years have astronomers discovered brown dwarfs in numbers sufficient to classify them into more than one category.

We can also easily determine the upper mass limit to star formation. A star with a mass greater than about a hundred times the

Sun's will have a luminosity so great—such an enormous out-
pouring of energy in the form of visible light, infrared, and ultraviolet—that any additional gas and dust attracted toward the star
will be pushed away by the intense pressure of starlight. The
star's photons push on the dust grains within the cloud, which in
turn carry the gas away with them. Here starlight couples irre-
versibly to dust. This radiation pressure operates so effectively
that just a few high-mass stars within a dark, obscuring cloud will
have luminosities sufficient to disperse nearly all its interstellar
matter, laying bare to the universe dozens, if not hundreds, of
brand-new stars—all siblings, really—for the rest of the galaxy
to see.

Whenever you gaze at the Orion nebula, located just below the
three bright stars of Orion's Belt, midway along the Hunter's
somewhat fainter sword, you can see a stellar nursery of just this
sort. Thousands of stars have been born within this nebula, while
thousands more await their birth, soon to create a giant star cluster that becomes more and more visible to the cosmos as the nebula dissipates. The most massive new stars, forming a group called
the Orion Trapezium, are busy blowing a giant hole in the middle of the cloud from which they formed. Hubble Telescope
images of this region reveal hundreds of new stars in this zone
alone, each infant swaddled within a nascent protoplanetary disk
made of dust and other molecules drawn from the original cloud.
And within each of these disks, a planetary system is forming.

Ten billion years after the Milky Way formed, star formation
continues today at multiple locations in our galaxy. Even though
most of the star formation that will ever occur in a typical giant
galaxy like ours has already taken place, we are fortunate that
new stars continue to form, and will do so for many billion years
to come. Our good fortune lies in our ability to study the forma-

tion process and the youngest stars, seeking clues that will reveal, in all its glory, the complete story of how stars pass from cold gas and dust to luminous maturity.

How old are the stars? No star wears its age on its sleeve, but some show their ages in their spectra. Among the various means that astrophysicists have devised to judge the ages of stars, spectra forms the most reliable hinge for analyzing the different colors of starlight in detail. Every color—every wavelength and frequency of the light waves we observe—tells a story about how matter made the starlight, or affected that light as it left the star, or happened to lie along the line of sight between ourselves and the star. Through close comparison with laboratory spectra, physicists have determined the multitude of ways that different types of atoms and molecules affect the rainbow of colors in visible light. They can apply this fertile knowledge to observations of stellar spectra, and deduce the numbers of atoms and molecules that have affected light from a particular star, as well as the temperature, pressure, and density of those particles. From years of comparing laboratory spectra with the spectra of stars, together with laboratory studies of the spectra of different atoms and molecules, astrophysicists have learned how to read an object's spectrum like a cosmic fingerprint, one that reveals what physical conditions exist within a star's outer layers, the region from which light streams directly outward into space. In addition, astrophysicists can determine how atoms and molecules floating in interstellar space at much cooler temperatures may have affected the spectrum of the starlight they observe, and can likewise deduce the chemical composition, temperature, density, and pressure of this interstellar matter.

In this spectral analysis, each different type of atom or molecule has its own story to tell. The presence of molecules of any type, for example, revealed by their characteristic effects on certain colors in the spectrum, demonstrates that the temperature in a star's outer layers must be less than about 3,000° Celsius (about 5,000° Fahrenheit). At higher temperatures, molecules move so rapidly that their collisions break them apart into individual atoms. By extending this type of analysis over many different substances, astrophysicists can derive a nearly complete picture of the detailed conditions in stellar atmospheres. Some hard-working astrophysicists are said to know far more about the spectra of stars they love than they do about their own families. This may have its down side for interpersonal relations even as it increases human understanding of the cosmos.

Of all nature's elements—of all the different types of atoms that can create patterns in a star's spectrum—astrophysicists recognize and use one in particular to find the ages of the youngest stars. That element is lithium, the third simplest and lightest in the periodic table, and familiar to some on Earth as the active ingredient of some antidepressant medications. In the periodic table of the elements, lithium occupies the position immediately after hydrogen and helium, which are deservedly far more famous because they exist in immensely greater amounts throughout the cosmos. During its first few minutes, the universe fused hydrogen into helium nuclei in great numbers, but made only relatively tiny amounts of any heavier nucleus. As a result, lithium remained a rather rare element, distinguished among astrophysicists by the cosmic fact that stars hardly ever make more lithium, but only destroy it. Lithium rides down a one-way street because every star has more effective nuclear fusion reactions to destroy lithium than to create it. As a result, the cosmic supply of lithium has steadily decreased and continues to do so. If you want some, now would be a good time to acquire it.

For astrophysicists, this simple fact about lithium makes it a highly useful tool for measuring the ages of stars. All stars begin their lives with their fair and proportionate share of lithium, left behind by the nuclear fusion that occurred during the universe's first half hour—and during the big bang itself. And what is that fair share? About one in every 100 billion nuclei. After a newborn star begins its life with this "richness" of lithium, things go downhill, lithiumwise, as nuclear reactions within the star's core slowly consume lithium nuclei. The steady and sometimes episodic mixing of matter in the core with matter outside carries material outward, so that after thousands of years, the star's outer layers can reflect what previously happened in its core.

When astrophysicists look for the youngest stars, they therefore follow a simple rule: Look for the stars with the *greatest* abundance of lithium. Each star's number of lithium nuclei in proportion to, for example, hydrogen (determined from careful study of the star's spectrum), will locate the star at some point along a graph that shows how stars' ages correlate with lithium in their outer layers. By using this method, astrophysicists can identify, with confidence, the youngest stars in a cluster, and can assign each of those stars a lithium-based age. Because stars are efficient destroyers of lithium, older stars show little if any of the stuff. Hence the method works well only for stars less than few hundred million years old. But for these younger stars, the lithium approach works wonders. A recent study of two dozen young stars in the Orion nebula, all of which have masses close to the Sun's, show ages that range between 1 and 10 million years. Some day astrophysicists may well identify still younger stars, but for now, 1 million years represents about the best they can do.

Except for dispersing the cocoons of gas from which they formed, groups of newborn stars bother nobody for a long time, as they quietly fuse hydrogen into helium in their cores and destroy their lithium nuclei as part of their fusion reactions. But nothing lasts forever. Over many million years, in response to the continual gravitational perturbations from enormous clouds that pass by, most would-be star clusters "evaporate," as its members scatter into the general pool of stars in the galaxy.

Nearly 5 billion years after our star formed, the identity of the Sun's siblings has vanished, whether or not those stars remain alive. Of all the stars in the Milky Way and other galaxies, those with low masses consume their fuel so slowly that they live practically forever. Intermediate-mass stars such as our Sun eventually turn into red giants, expanding their outer gas layers a hundredfold in size as they slide toward death. These outer layers become so tenuously connected to the star that they drift into space, exposing a core of spent nuclear fuels that powered the stars' 10-billion-year lives. The gas that returns to space will be swept up by passing clouds, to participate in later rounds of star formation.

Despite their rarity, the highest-mass stars hold nearly all the evolutionary cards. Their high masses give them the greatest stellar luminosities—some of them can boast a million times the Sun's—and because they consume their nuclear fuel far more rapidly than low-mass stars do, they have the shortest lives of all stars, only a few million years, or even less. Continued thermonuclear fusion within high-mass stars allows them to manufacture dozens of elements in their cores, starting with hydrogen and proceeding to helium, carbon, nitrogen, oxygen, neon, magnesium, silicon, calcium, and so on, all the way to iron. These stars forge still more elements in their final fires, which can briefly outshine a star's entire home galaxy. Astrophysicists call each of these outbursts a supernova, similar in appearance (though quite different in their origin) to the Type Ia supernovae described in Chapter 5. A supernova's explosive energy spreads both the previously made and the freshly minted elements through the galaxy, blowing holes in its distribution of gas and enriching nearby clouds with the raw materials to make new dust grains. The blast moves supersonically through these interstellar clouds, compressing their gas and dust, possibly creating some of the high-density pockets needed to form stars.

The greatest gift to the cosmos from these supernovae consists of all the elements other than hydrogen and helium—elements capable of forming planets and protists and people. We on Earth live on the product of countless stars that exploded billions of years ago, in epochs of Milky Way history long before our Sun and its planets, condensing within the dark and dusty recesses of an interstellar cloud—itself endowed with chemical enrichment furnished from previous generations of high-mass stars.

How did we come to taste this delicious kernel of knowledge, the fact that all the elements beyond helium were forged within stars? The authors' award for the most underappreciated scientific discovery of the twentieth century goes to the recognition that supernovae—the explosive death throes of high-mass stars—provide the primary source for the origin and abundances of heavy elements in the universe. This relatively unsung realization appeared in a lengthy research article, published in 1957 in the U.S. journal *Reviews of Modern Physics* under the title "The Synthesis of the Elements in Stars," and written by E. Margaret Burbidge, Geoffrey R. Burbidge, William Fowler, and Fred Hoyle. In this paper, the four scientists created a theoretical and computational framework that freshly interpreted and melded together forty years of musings by other scientists on two key topics: the sources of stellar energy and the transmutation of chemical elements.

Cosmic nuclear chemistry, the quest to understand how nuclear fusion makes and destroys different types of nuclei, has always been a messy business. The crucial questions have always included: How do the various elements behave when various temperatures and pressures act upon them? Do the elements fuse or do they split? How easily do they do this? Do these processes lib-

erate new kinetic energy or absorb existing kinetic energy? And
how do the processes differ for each element in the periodic table?

What does the periodic table of the elements mean to you? If
you are like most former students, you will remember a giant chart
on the wall of your science class, tricked out with mysterious boxes
in which cryptic letters and symbols murmured tales of dusty laboratories to be avoided by young souls in transition. But to those
who know its secrets, this chart tells a hundred stories of cosmic
violence that brought its components into existence. The periodic
table lists every known element in the universe, arranged by the
increasing number of protons in each element's nuclei. The two
lightest elements are hydrogen, with one proton per nucleus, and
helium, with two. As the four authors of the 1957 paper saw, under
the right conditions of temperature, density, and pressure, a star
can use hydrogen and helium to create all the other elements in
the periodic table.

The details of this creation process, and of other interactions
that destroy nuclei rather than create them, provide the subject
matter for nuclear chemistry, which involves the calculation and
use of "collision cross sections" to measure how closely one particle must approach another before they are likely to interact significantly. Physicists can easily calculate collision cross sections for
cement mixers, or double-wide mobile homes moving down the
street on flatbed trucks, but they face greater challenges in analyzing the behavior of tiny, elusive subatomic particles. A detailed
understanding of collision cross sections enables physicists to predict nuclear reaction rates and pathways. Often small uncertainties in their tables of cross sections lead them into wildly
erroneous conclusions. Their difficulties resemble what would
happen if you tried to navigate your way through one city's subway system with another city's subway map as your guide: your
basic theory would be correct, but the details could kill you.

Despite their ignorance of accurate collision cross sections, scientists during the first half of the twentieth century had long sus-

pected that if exotic nuclear processes exist anywhere in the universe, the centers of stars seemed likely places to find them. In 1920, the British theoretical astrophysicist Sir Arthur Eddington published a paper entitled the "The Internal Constitution of the Stars," in which he argued that the Cavendish Laboratory in England, the leading center for atomic and nuclear physics research, could not be the only place in the universe that managed to change some elements into others:

> But is it possible to admit that such a transmutation is occurring? It is difficult to assert, but perhaps more difficult to deny, that this is going on . . . and what is possible in the Cavendish Laboratory may not be too difficult in the sun. I think that the suspicion has been generally entertained that the stars are the crucibles in which the lighter atoms which abound in the nebulæ are compounded into more complex elements.

Eddington's paper, which foreshadowed the detailed research of Burbidge, Burbidge, Fowler, and Hoyle, appeared several years before the discovery of quantum mechanics, without which our understanding of the physics of atoms and nuclei must be judged feeble at best. With remarkable prescience, Eddington began to formulate a scenario for star-generated energy via the thermonuclear fusion of hydrogen to helium and beyond:

> We need not bind ourselves to the formation of helium from hydrogen as the sole reaction which supplies the energy [to a star], although it would seem that the further stages in building up the elements involve much less liberation, and sometimes even absorption, of energy. The position may be summarised in these terms: the atoms of all elements are built of hydrogen atoms bound together, and presumably have at one time been formed from hydrogen; the interior of a star seems as likely a place as any for the evolution to have occurred.

Any model of the transmutation of the elements ought to explain the observed mix of elements found on Earth and elsewhere in the universe. To do this, physicists needed to find the fundamental process with which stars generate energy by turning one element into another. By 1931, with theories of quantum mechanics rather well developed (although the neutron had not yet been discovered), the British astrophysicist Robert d'Escourt Atkinson published an extensive paper, summarized as a "synthesis theory of stellar energy and of the origin of the elements . . . in which the various chemical elements are built up step by step from the lighter ones in stellar interiors, by the successive incorporation of protons and electrons one at a time."

In the same year, the American nuclear chemist William D. Harkins published a paper noting that "elements of low atomic weight [the number of protons plus neutrons in each nucleus] are more abundant than those of high atomic weight and that, on the average, the elements with even atomic numbers [the numbers of protons in each atomic nucleus] are about 10 times more abundant than those with odd atomic numbers of similar value." Harkins surmised that the relative abundances of the elements depend on nuclear fusion rather than on chemical processes such as combustion, and that the heavy elements must have been synthesized from the light ones.

The detailed mechanism of nuclear fusion in stars could ultimately explain the cosmic presence of many elements, especially those that you will obtain each time you add the two-proton, two-neutron helium nucleus to your previously forged element. These constitute the abundant elements with "even atomic numbers" that Harkins described. But the existence and relative numbers of many other elements remained unexplained. Some other means of element buildup must have been at work in the cosmos.

The neutron, discovered in 1932 by the British physicist James Chadwick while working at the Cavendish Laboratories, plays a significant role in nuclear fusion that Eddington could not have imagined. To assemble protons requires hard work, because protons naturally repel one another, as do all particles with the same sign of electric charge. To fuse protons, you must bring them sufficiently close (often by way of high temperatures, pressures, and densities) to overcome their mutual repulsion for the strong nuclear force to bind them together. The chargeless neutron, however, repels no other particle, so it can simply march into somebody else's nucleus and join the other assembled particles, held there by the same force that binds the protons. This step does not create another element, which is defined by a different number of *protons* in each nucleus. By adding a neutron, we make an "isotope" of the nucleus of the original element, which differs only in detail from the original nucleus because its total electric charge remains unchanged. For some elements, the freshly captured neutron proves to be unstable once it joins the nucleus. In that case, the neutron spontaneously converts itself into a proton (which stays put in the nucleus), and an electron (which escapes immediately). In this way, like the Greek soldiers who breached the walls of Troy by hiding inside a wooden horse, protons can sneak into a nucleus in the guise of neutrons.

If the ongoing flow of neutrons stays high, each nucleus can absorb many neutrons before the first one decays. These rapidly absorbed neutrons help to create an ensemble of elements whose origin is identified with the "rapid neutron capture process," and differ from the assortment of elements that result when neutrons are captured slowly, where each successive neutron decays into a proton before the nucleus captures the next one.

Both the rapid and the slow neutron capture processes are responsible for creating many of the elements not otherwise formed through traditional thermonuclear fusion. The remaining elements in nature can be made by a few other processes, including slamming high-energy photons (gamma rays) into the nuclei of heavy atoms, which then break apart into smaller ones.

At the risk of oversimplifying the life cycle of a high-mass star, we may state that each star lives by generating and releasing the energy in its interior that allows the star to support itself against gravity. Without its production of energy through thermonuclear fusion, each stellar ball of gas would simply collapse under its own weight. This fate weighs on stars that exhaust their supplies of hydrogen nuclei (protons) in their cores. As already noted, after converting its hydrogen into helium, the core of a massive star will next fuse helium into carbon, then carbon to oxygen, oxygen to neon, and so forth up to iron. To successively fuse this sequence of heavier and heavier elements requires successively higher temperatures for the nuclei to overcome their natural repulsion. Fortunately this happens all by itself, because at the end of each intermediate stage, when the star's energy source temporarily shuts off, the inner regions collapse, the temperature rises, and the next pathway of fusion kicks in. Since nothing lasts forever, the star eventually confronts one enormous problem: The fusion of iron does not release energy, but instead absorbs it. This brings bad news to the star, which can now no longer support itself against gravity by pulling a new energy-releasing process out of its nuclear fusion hat. At this point, the star suddenly collapses, forcing its internal temperature to rise so rapidly that a gigantic explosion ensues as the star blows its guts to smithereens.

Throughout each explosion, the availability of neutrons, protons, and energy allows the supernova to create elements in many different ways. In their 1957 article, Burbidge, Burbidge, Fowler, and Hoyle combined (1) the well-tested tenets of quantum mechanics; (2) the physics of explosions; (3) the latest collision cross sections; (4) the varied processes that transmute elements into one another;

and (5) the basics of stellar evolutionary theory to implicate supernova explosions decisively as the primary source of all the elements heavier than hydrogen and helium in the universe.

With high-mass stars as the source of heavy elements, and supernovae as the smoking gun of element distribution, the fab four acquired the solution to one other problem for free: when you forge elements heavier than hydrogen and helium in stellar cores, you do the rest of the universe no good unless you somehow cast those elements forth into interstellar space, making them available to form worlds with wombats. Burbidge, Burbidge, Fowler, and Hoyle unified our understanding of nuclear fusion in stars with the element production visible throughout the universe. Their conclusions have survived decades of skeptical analysis, so their publication stands as a turning point in our knowledge of how the universe works.

Yes, Earth and all its life comes from stardust. No, we have not solved all of our cosmic chemical questions. A curious contemporary mystery involves the element technetium, which, in 1937, was the first element to be created artificially in Earthbound laboratories. (The word "technetium," along with others that use the prefix "tech-," derive from the Greek *technetos*, which translates to "artificial.") We have yet to discover technetium on Earth, but astronomers have found it in the atmospheres of a small fraction of the red giant stars in our galaxy. This would hardly surprise us, were it not for the fact that technetium decays to form other elements, and does so with a half-life of a mere 2 million years, far shorter than the age and life expectancy of the stars in which we observe it. This conundrum has led to exotic theories that have yet to achieve consensus within the community of astrophysicists.

Red giants with these peculiar chemical properties are rare, but sufficiently nettlesome for a cadre of astrophysicists (mostly spectroscopists) who specialize in the subject to generate and distribute the *Newsletter of Chemically Peculiar Red Giant Stars*. Not avail-

able on most newsstands, this publication typically contains conference news and updates on research still in progress. To interested scientists, these ongoing chemical mysteries have an allure as
strong as the questions related to black holes, quasars, and the early
universe. But you hardly ever read about them. Why? Because,
quite typically, the media has predetermined what deserves cover-
age and what does not. Apparently the news about the cosmic origins of every element in your body and your planet doesn't make
the cut.

Here is your chance to redress the wrongs that contemporary
society has inflicted upon you. Let's take a journey through the
periodic table, pausing here and there to note the most intriguing
facts about the various elements, and to admire how the cosmos
made them all from the hydrogen and helium that emerged from
the big bang.

CHAPTER 10

The Elemental Zoo

The periodic table of the elements, lovingly created by chemists and physicists during the past two centuries, embodies organizing principles that explain the chemical behavior of all the elements that we know in the universe, or may someday discover. For this reason, we ought to regard the periodic table as a cultural icon, an exemplar of our society's ability to organize its knowledge. The table testifies to the enterprise of science as an international human adventure, conducted not only in laboratories but also in particle accelerators, and at the space and time frontiers of the entire cosmos.

Amid this well-merited respect, every now and then an entry in the periodic table will strike even a grown-up scientist as a strange beast in a zoo of one-of-a-kind animals conceived and executed by Dr. Seuss. How else can we believe that sodium is a deadly, reactive metal that you can cut with a butterknife, and that pure chlorine is an evil-smelling, deadly gas—yet when we combine sodium and chlorine, we make sodium chloride, a harmless compound essential to life, better known as table salt? What about hydrogen and oxygen, two of the most abundant elements on Earth and in the universe? One is an explosive gas, while the other promotes violent combustion; yet adding the two produces liquid water, which puts out fires.

Amid all the chemical interactions in the periodic table's little shop of possibilities, we find the elements most significant to the cosmos. These offer the chance to view the table through the lens of an astrophysicist. We shall grasp that chance and dance our way across the table, saluting its most distinguished entries and admiring its little oddities.

The periodic table emphasizes the fact that each of nature's elements distinguishes itself from all others by its "atomic number," the number of protons (positive electric charges) in each nucleus of that element. Complete atoms always have a number of electrons (negative electric charges) orbiting the nucleus equal to the element's atomic number, so the total atom has zero electric charge. Different isotopes of a particular element have the same number of protons and electrons, but different numbers of neutrons.

Hydrogen, with only one proton in each nucleus, is the lightest and simplest element, made entirely during the first few minutes after the big bang. Out of the ninety-four naturally occurring elements, hydrogen claims more than two thirds of all the atoms in human bodies and more than 90 percent of all the atoms in the cosmos, including the Sun and its giant planets. The hydrogen inside the core of the Sun's most massive planet, Jupiter, feels so much pressure from the overlying layers that it behaves more like an electromagnetically conductive metal than a gas, and helps to create the strongest magnetic field among the Sun's planets. The English chemist Henry Cavendish discovered hydrogen in 1766 while experimenting with H_2O (*hydro-genes* is the Greek word for water-forming, whose *gen* appears in such English words as "genetic"), though his fame among astronomers rests on his having been the first person to calculate Earth's mass accurately by measuring the gravitational constant G that appears in Newton's famous equation for gravity. Every second of every day and night, 4.5 billion tons of fast-moving hydrogen nuclei (protons) slam together to make helium nuclei within the Sun's 15-million-degree (Celsius) core. About 1 percent of the mass involved in this fusion transforms itself into energy, leaving the other 99 percent in the form of helium.

Helium, the second most abundant element in the universe, can be found on Earth only in a few underground pockets that trap this gas. Most of us know only helium's whimsical side, available for testing through over-the-counter purchases. When you inhale helium, its low density in comparison with atmospheric gases increases the vibrational frequency within your windpipe, causing you to sound like Mickey Mouse. The cosmos contains four times more helium than all other elements combined (not counting hydrogen). One of the pillars of big bang cosmology is the prediction that throughout the cosmos, no fewer than about 8 percent of all atoms are helium, which the well-mixed primeval fireball manufactured during its immediate post-birth pangs. Since the thermonuclear fusion of hydrogen within stars produces additional helium, some regions of the cosmos can accumulate more than their initial 8 percent share of helium, but—just as the big bang model predicts—no one has ever found a region of our galaxy or anybody else's galaxy with less.

Some thirty years before they discovered and isolated helium on Earth, astrophysicists had detected helium in the Sun by the telltale features that they saw in the Sun's spectrum of light during the total eclipse of 1868. They naturally named this previously unknown material helium after Helios, the Greek sun god. With 92 percent of hydrogen's buoyancy in air, but without the explosive characteristics of hydrogen that destroyed the German

Hindenburg dirigible, helium provides the gas of choice for the outsized balloon characters of the Macy's Thanksgiving Day parade, making the famed department store second only to the U.S. military as the world's top consumer of helium.

Lithium, the third simplest element in the universe, has three protons in each nucleus. Like hydrogen and helium, lithium was made soon after the big bang, but unlike helium, which is often made in subsequent nuclear reactions, lithium will be *destroyed* by every nuclear reaction that occurs in stars. Hence we expect to find no object or region with lithium present in more than the relatively small relative abundances—no more than 0.0001 percent of the total—produced in the early universe. As predicted by our model of element formation during the first half hour, no one has yet found a galaxy with more lithium than this upper limit. The combination of the upper limit on helium and the lower limit on lithium furnishes us with a potent dual constraint to apply in testing the theory of big bang cosmology. A similar test of the big bang model of the universe, which it has passed with flying colors, compares the abundance of deuterium nuclei, each of which has one proton and one neutron, with the amount of ordinary hydrogen. Fusion during the first few minutes produced both of these nuclei, but made far more of simple hydrogen (just one proton).

Like lithium, the next two elements in the periodic table, **beryllium** and **boron** (with four and five protons, respectively, in each nucleus) owe their origin mainly to thermonuclear fusion in the early universe, and they appear only in relatively modest numbers throughout the cosmos. The scarcity on Earth of the three lightest elements after hydrogen and helium makes them bad news for those who accidentally ingest them, since evolution has proceeded essentially without encountering them. Intriguingly, controlled doses of lithium do seem to relieve certain types of mental illness.

With **carbon**, element number six, the periodic table springs into glorious efflorescence. Carbon atoms, with six protons in every nucleus, appear in more kinds of molecules than the sum of all non-carbon-containing molecules combined. The cosmic abundance of carbon nuclei—forged in the cores of stars, churned to their surfaces, and released in copious amounts into the Milky Way galaxy—joins with carbon's ease in forming chemical combinations to make carbon the best element on which to base the chemistry and diversity of life. Just edging out carbon in abundance, **oxygen** (eight protons per nucleus) also offers a highly reactive and abundant element, similarly forged within and released from aging stars and stars that explode as supernovae. Both oxygen and carbon constitute major ingredients for life as we know it. The same processes made and distributed **nitrogen**, element number seven, which again appears in great quantities throughout the universe.

But what about life as we don't know it? Could other life forms use a different element as the heart of their complex shapes? How about life based on **silicon**, element number 14? Silicon sits directly below carbon on the periodic table, which means (see how useful the table can be to those who know its secrets) that silicon can create the same sorts of chemical compounds that carbon does, with silicon taking the place of carbon. In the end, we expect carbon to prove superior to silicon, not only because carbon has ten times the abundance of silicon in the cosmos but also because silicon forms chemical bonds that are either substantially stronger or noticeably weaker than those that carbon makes. In particular, the strength of the bonds between silicon and oxygen makes tough rocks, whereas complex molecules based on silicon lack the hardiness to survive ecological stresses that carbon-based atoms exhibit. These facts don't stop science fiction writers from championing silicon, thus keeping exobiological speculation on its toes and allowing us to wonder what the first truly alien life form will be like.

In addition to forming an active ingredient in table salt, **sodium** (eleven protons per nucleus) glows across this great land as hot sodium gas in most municipal street lamps. These lamps "burn" brighter, longer, and use less energy than conventional incandescent bulbs do. They come in two varieties: the common high-pressure lamps, which look yellow-white, and the rarer, low-pressure lamps, which look orange. It turns out that while all light pollution hurts astronomy, low-pressure sodium lamps inflict less harm because their contamination, much more narrowly confined in color, can be easily accounted for and removed from telescope data. In a model of town-telescope cooperation, the entire city of Tucson, Arizona, the closest large municipality to the Kitt Peak National Observatory, has, by agreement with the local astronomers, converted all its streetlights to low-pressure sodium lamps—which also turn out to be more efficient, and therefore save energy for the city.

Aluminum (twelve protons per nucleus) provides nearly 10 percent of Earth's crust, yet remained unknown to the ancients and unfamiliar to our grandparents because it combines so effectively with other elements. Its isolation and identification occurred only in 1827, and aluminum did not enter common household use until the late 1960s, when tin cans and tin foil yielded to aluminum cans and aluminum foil. Because polished aluminum makes a near-perfect reflector of visible light, astronomers today coat nearly all their telescope mirrors with a thin film of aluminum atoms.

Although **titanium** (thirteen protons per nucleus) has a density 70 percent greater than aluminum's, it's more than twice as strong. Its strength and relative lightness make titanium—the ninth most abundant element in Earth's crust—a modern darling for many applications, such as military aircraft components, that require a light, strong metal.

In most cosmic locations, oxygen atoms outnumber carbon. In

stars, once every carbon atom has latched onto one of the available oxygen atoms to form carbon monoxide or carbon dioxide molecules, the leftover oxygen atoms bond with other elements, such as titanium. The spectra of the light from red-giant stars are riddled with features created by titanium oxide (molecules of TiO), which itself is no stranger to stars on Earth: star sapphires and rubies owe their radiant asterisms to titanium oxide impurities within their crystal lattices, with aluminum oxide impurities adding extra color. Furthermore, the white paint used for telescope domes features titanium oxide, which happens to radiate infrared with high efficiency, a fact that greatly reduces the daytime heat accumulated within the dome. At nightfall, with the dome open, the air temperature near the telescope falls more rapidly to the temperature of the nighttime air, reducing atmospheric refraction and allowing the light from stars and other cosmic objects to arrive with greater sharpness and clarity. Although not directly named for a cosmic object, titanium derives its handle from the Titans of Greek mythology, as does Titan, Saturn's largest moon.

Carbon may be the most significant element in life, but by many measures, **iron**, element number 26, ranks as the most important of all the elements in the universe. Massive stars manufacture elements in their core, marching through the periodic table in the sequence of increasing number of protons per nucleus, from helium to carbon to oxygen to neon, and so forward all the way to iron. With twenty-six protons and at least as many neutrons in its nucleus, iron has a distinctive quality that derives from the quantum mechanics rules that govern how protons and neutrons interact: Iron nuclei have the highest binding energy per nuclear particle (proton or neutron). This means something quite simple. If you seek to split iron nuclei (in what physicists call "fission"), you must provide them with additional energy. On the other hand, if you combine iron atoms (a process called "fusion"), they will also absorb energy. It takes energy to fuse iron nuclei and it takes energy to split them apart. For all other elements, only one or the other half of this dual description applies.

Stars, however, are in the business of using $E = mc^2$ to turn mass into energy, which they must do to oppose their tendency to collapse under their own gravity. When stars fuse nuclei in their cores, nature demands, and obtains, nuclear fusion that releases energy. By the time that a massive star fuses most of the nuclei in its core into iron, it has exhausted all its options for using thermonuclear fusion to generate energy, because any further fusion will require rather than release energy. Deprived of a source of energy from thermonuclear fusion, the star's core will collapse under its own weight, then instantly rebound in a titanic explosion known as a supernova, outshining a billion suns for more than a week. Such supernovae occur because of the special property of iron nuclei—their refusal either to fuse or to split without an input of energy.

By describing hydrogen, helium; lithium, beryllium, and boron; carbon, nitrogen, and oxygen; and aluminum, titanium, and iron, we have surveyed nearly all of the key elements that make the cosmos—and life on Earth—go round.

Simply for the cosmic hell of it, let's have a quick look at some far more obscure entries on the periodic table. You will almost certainly never own any significant quantities of these elements, but scientists find them not only intriguing riffs on nature's bounty but also highly useful in special circumstances. Consider, for example, the soft metal **gallium** (thirty-one protons per nucleus). Gallium has such a low melting point that the heat from your hand will make it liquefy. Apart from this parlor demo opportunity, gallium provides astrophysicists with the active ingredient in gallium chloride, a variant on table salt (sodium chloride) that proves valuable in experiments that detect neutrinos from the Sun's core. To capture these elusive neutrinos, astrophysicists create a 100-ton vat of liquid gallium chloride and set it deep underground (to screen out effects from less penetrating particles), then watch it carefully for the results of any collisions between the neutrinos and the gallium nuclei, which turn the nuclei into germanium nuclei, each of which has thirty-two protons. Every transformation of gallium into germanium produces X-ray photons, which can be detected and measured every time that a nucleus gets slammed. By using these gallium-chloride "neutrino telescopes," astrophysicists resolved what they had called the "solar neutrino problem," the fact that earlier types of neutrino detectors found neutrinos in smaller numbers than the theory of thermonuclear fusion in the Sun's core had predicted.

Every nucleus of the element **technetium** (atomic number 43) is radioactive, decaying after a few moments or a few million years into other types of nuclei. Not surprisingly, we find technetium nowhere on Earth except in particle accelerators, where we make it on demand. For reasons not yet fully understood, technetium lives in the atmospheres of a select subset of red giant stars. As we noted in the previous chapter, this would cause astrophysicists no alarm—except that technetium has a half-life of a mere 2 million years, far, far shorter than the ages and life expectancies of the stars in which we find it. This proves that the stars cannot have been born with the stuff, for if they had been, none would be left by now. Astrophysicists also lack any known mechanism to create technetium in a star's core *and* to have it dredge itself up to the surface where they observe it, a matter of uneasy fact that has spawned exotic explanations, still shy of consensus within the astrophysics community.

Along with osmium and platinum, **iridium** gives us one of the three densest elements on the periodic table—two cubic feet of iridium (atomic number 77) weighs as much as a Buick, which makes it one of the world's best paperweights, able to defy all known office fans and window breezes. Iridium also gives scientists the world's most famous smoking gun. All over the world, a thin layer of iridium-rich material appears at the geological layer that marks the famous K-T boundary, laid down 65 million years ago. Not coincidentally, most biologists believe, that boundary also marks the time when every land species larger than a breadbox, including the legendary dinosaurs, went extinct. Iridium is rare on Earth's surface, but ten times more common in metallic asteroids. Whatever might have been your favorite theory for destroying the dinosaurs, a ten-mile-wide killer asteroid from outer space, capable of raising a worldwide blanket of light-blocking debris before slowly raining downward several months later, now seems quite compelling.

It's not clear how Albert would have felt about this, but physicists discovered a previously unknown element in the debris from the first hydrogen bomb test in the Pacific (November 1952) and named it **einsteinium** in his honor. Armageddium might have been more suitable.

While helium derives its name from the Sun itself, ten other elements in the periodic table draw their names from objects that orbit the Sun:

Phosphorus, which means "light-bearing" in Greek, was the ancient name for the planet Venus when it appeared before sunrise in the dawn sky.

Selenium comes from *selene*, the Greek word for the Moon, so named because this element was always found in association with the element tellurium, which had already been named for Earth, from the Latin *tellus*.

On January 1, 1801, the first day of the nineteenth century, the Italian astronomer Giuseppe Piazzi discovered a new planet orbiting the Sun within the suspiciously large gap between Mars and Jupiter. Maintaining the tradition of naming planets after Roman gods, Piazzi called the object Ceres after the goddess of harvest, which also provides the root for our word "cereal." The excitement in the scientific community over Piazzi's find caused the next element to be discovered to be named **cerium** in its honor. Two years later, another planet was found, orbiting the Sun within the same gap as Ceres. This object received the name Pallas, from the Roman goddess of wisdom; like cerium before it, the next element discovered thereafter was named **palladium** in its honor. The naming party ended a few decades later, after dozens more of these planets were discovered in much the same location, and after closer analysis revealed that these objects are much, much smaller than the smallest known planets. A new swath of real estate had come into view within the solar system, consisting of small, craggy chunks of rock and metal. Ceres and Pallas turned out to be not planets but asteroids, objects only a few hundred miles across. They live in the asteroid belt, now known to contain millions of objects, of which astronomers have catalogued and named upward of fifteen thousand—somewhat more than the number of elements in the periodic table.

The metal **mercury**, which assumes a viscous liquid form at room temperature, owes its name to the speedy Roman messenger god. So too does the planet Mercury, the fastest-moving of all the planets in the solar system.

Thorium's name comes from Thor, the hammer-and-thunder-wielding Scandinavian god, who corresponds to the lightning-bolt-wielding Jupiter in Roman mythology. By jove, recent Hubble Space Telescope images of Jupiter's polar regions reveal extensive electrical discharges deep within its turbulent cloud layers.

Saturn, most people's favorite planet, has no element named for it, but Uranus, Neptune, and Pluto are famously represented. The element **uranium**, discovered in 1789, received its name in honor of William Herschel's planet, discovered by him just eight years earlier. All isotopes of uranium are unstable, spontaneously but slowly decaying to lighter elements, a process accompanied by the release of energy. If you can arrange to speed up the rate of decay with a "chain reaction" among uranium nuclei, you have the explosive energy release required for a bomb. In 1945, the United States exploded the first uranium bomb (familiarly called an atomic bomb or A-bomb) to be used in warfare, incinerating the Japanese city of Hiroshima. With ninety-two protons packed in each nucleus, uranium wins the prize as the largest and heaviest element to occur naturally, although trace amounts of still larger and heavier elements appear in places where uranium ore is mined.

If Uranus merited an element, so did Neptune. Unlike uranium, however, which was identified soon after its planet, neptunium was discovered in 1940 in the particle accelerator called the Berkeley Cyclotron, ninety-seven years after the German astronomer John Galle found Neptune in a spot on the sky predicted as the most likely spot by the French mathematician Joseph Le Verrier, who studied Uranus' unexplained orbital behavior and deduced the existence of a farther planet. Just as Neptune comes immediately after Uranus in the solar system, neptunium comes right after uranium in the periodic table of the elements.

Particle physicists working at the Berkeley cyclotron discovered more than half a dozen elements not found in nature, including **plutonium**, which immediately follows neptunium in the periodic table and bears the name of Pluto, which the young astronomer Clyde Tombaugh found in 1930 on photographs taken at Arizona's Lowell Observatory. As with the discovery of Ceres 129 years earlier, excitement ran high. Pluto was the first planet discovered by an American and, in the absence of accurate observational data, was widely believed to be a planet of size and mass commensurate with those of Uranus and Neptune. As our measurements of Pluto's size improved, Pluto kept getting smaller. Our knowledge of Pluto's dimensions did not stabilize until the late 1970s, during the *Voyager* missions to the outer solar system. We now know that cold, icy Pluto is by far the Sun's smallest planet, with the embarrassing distinction of being smaller than the solar system's six largest moons. As with the asteroids, astronomers later found hundreds of other objects in similar locations, in this case in the outer solar system with orbits similar to Pluto's. These objects signaled the existence of a heretofore undocumented reservoir of small icy objects, now called the Kuiper Belt of comets. A purist could argue that like Ceres and Pallas, Pluto slipped into the periodic table under false pretenses.

Like uranium nuclei, plutonium nuclei are radioactive. These nuclei formed the active ingredient in the atomic bomb dropped on the Japanese city of Nagasaki, just three days after the uranium bombing of Hiroshima, bringing a swift end to World War II. Scientists can use small quantities of plutonium, which produces energy at a modest, steady rate, to power radioisotope thermoelectric generators (abbreviated as RTGs) for spacecraft that travel to the outer solar system, where the intensity of sunlight falls below the level usable by solar panels. One pound of this plutonium will generate 10 million kilowatt-hours of heat energy, sufficient to power a household light bulb for eleven thousand years, or a human being for just about as long. Still drawing on their plutonium power to send messages to Earth, the two *Voyager* spacecraft launched in 1977 have now traveled far beyond Pluto's orbit. One of them, at nearly one hundred times Earth's distance from the Sun, has begun to enter true interstellar space by leaving the bubble that the Sun's outflow of electrically charged particles creates.

And so we end our cosmic journey through the periodic table of the chemical elements, right at the edge of the solar system. For reasons we have yet to determine, many people don't like chemicals, which may explain the perennial movement to rid foods of them. Perhaps sesquipedalian chemical names just sound dangerous. But in that case we should blame the chemists, and not the chemicals. Personally, we are quite comfortable with chemicals. Our favorite stars, as well as our best friends, are made of them.

Part IV

The Origin of Planets

CHAPTER 11

When Worlds Were Young

I n our attempts to uncover the history of the cosmos, we have continually discovered that the segments most deeply shrouded in mystery are those that deal with *origins*—of the universe itself, of its most massive structures (galaxies and galaxy clusters), and of the stars that provide most of the light in the cosmos. Each of these origin stories fills a vital role, not only in explaining how an apparently formless cosmos produced complex assemblages of different types of objects but also in determining how and why, 14 billion years after the big bang, we now find ourselves alive on Earth to ask, How did this all happen?

These mysteries arise in large part because during the cosmic "dark ages," when matter was just beginning to organize itself into self-contained units such as stars and galaxies, most of this matter generated little or no detectable radiation. The dark ages have left us with only the barest possibilities, still imperfectly explored, for observing matter during its early stages of organization. This in turn implies that we must rely, to an uneasily large extent, on our theories of how matter ought to behave, with relatively few points at which we can check these theories against observational data.

When we turn to the origin of planets, the mysteries deepen. We lack not only *observations* of the crucial, initial stages of planetary formation but also successful *theories* of how the planets began to form. To celebrate the positive, we note that the question, What made the planets?, has grown considerably broader in recent years. Throughout most of the twentieth century, this question centered on the Sun's family of planets. During the past decade, having discovered more than a hundred "exosolar" planets around relatively nearby stars, astrophysicists have acquired significantly more data from which to deduce the early history of planets, and in particular to determine how these astronomically small, dark, and dense objects formed along with the stars that give them light and life.

Astrnophysicists may now have more data, but they have no better answers than before. Indeed, the discovery of exosolar planets, many of which move in orbits far different from those of the Sun's planets, has in many ways confused the issue, leaving the story of planet formation no closer to closure. In simple summary, we can state that no good explanation exists of how the planets *began* to build themselves from gas and dust, though we can easily perceive how the formation process, once well underway, made larger objects from smaller ones, and did so within a rather brief span of time.

The beginnings of planet building pose a remarkably intractable problem, to the point that one of the world's experts on the subject, Scott Tremaine of Princeton University, has elucidated (partly in jest) Tremaine's laws of planet formation. The first of these laws states that "all theoretical predictions about the properties of exosolar planets are wrong," and the second that "the most secure prediction about planet formation is that it can't happen." Tremaine's humor underscores the ineluctable fact that planets do exist, despite our inability to explain this astronomical enigma.

More than two centuries ago, attempting to explain the formation of the Sun and its planets, Immanuel Kant proposed a "nebular hypothesis," according to which a swirling mass of gas and dust that surrounded our star-in-formation condensed into clumps that became the planets. In its broad outlines, Kant's hypothesis remains the basis for modern astronomical approaches to planet formation, having triumphed over the concept, much in vogue during the first half of the twentieth century, that the Sun's planets arose from a close passage of another star by the Sun. In that scenario, the gravitational forces between the stars would have drawn masses of gas from each of them, and some of this gas could then have cooled and condensed to form the planets. This hypothesis, promoted by the famed British astrophysicist James Jeans, had the defect (or the appeal, for those inclined in that direction) of making planetary systems extremely rare, because sufficiently close encounters between stars probably occur only a few times during the lifetime of an entire galaxy. Once astronomers calculated that almost all the gas pulled from the stars would evaporate rather than condense, they abandoned Jeans's hypothesis and returned to Kant's, which implies that many, if not most, stars should have planets in orbit around them.

Astrophysicists now have good evidence that stars form, not one by one but by the thousands and tens of thousands, within giant clouds of gas and dust that may eventually give birth to about a million individual stars. One of these giant stellar nurseries has produced the Orion nebula, the closest large star-forming region to the solar system. Within a few million years, this region will have produced hundreds of thousands of new stars, which will blow most of the nebula's remaining gas and dust into space, so that astronomers a hundred thousand generations from now will observe the young stars unencumbered by the remnants of their starbirthing cocoons.

Astrophysicists now use radio telescopes to map the distribution of cool gas and dust in the immediate vicinities of young stars. Their maps typically show that young stars do not sail through space devoid of all surrounding matter; instead, the stars usually have orbiting disks of matter, similar in size to the solar system, but made of hydrogen gas (and of other gases in lesser abundances) sprinkled throughout with dust particles. The term "dust" describes groups of particles that each contain several million atoms and have sizes much smaller than that of the period that ends this sentence. Many of these dust grains consist primarily of carbon atoms, linked together to form graphite (the chief constituent of the "lead" in a pencil). Others are mixtures of silicon and oxygen atoms—in essence tiny rocks, with mantles of ice surrounding their stony cores.

The formation of these dust particles in interstellar space has its own mysteries and detailed theories, which we may skip past with the happy thought that the cosmos *is* dusty. To make this dust, atoms have come together by the millions; in view of the extremely low densities between the stars, the likeliest sites for this process seem to be the extended outer atmospheres of cool stars, which gently blow material into space.

The production of interstellar dust particles provides an essential first step on the road to planets. This holds true not only for solid planets like our own but also for gas-giant planets, typified in the Sun's family by Jupiter and Saturn. Even though these planets consist primarily of hydrogen and helium, astrophysicists have concluded from their calculations of the planets' internal structure, along with their measurements of the planets' masses, that the gas giants must have solid cores. Of Jupiter's total mass, 318 times Earth's, several dozen Earth masses reside in a solid core. Saturn, with ninety-five times Earth's mass, also has a solid core with one or two dozen times the mass of Earth. The Sun's two smaller gas-giant planets, Uranus and Neptune, have proportionately larger solid cores. In these planets, with fifteen and seventeen times Earth's mass, respectively, the core may contain more than half of the planet's mass.

For all four of these planets, and presumably for all of the giant planets recently discovered around other stars, the planetary cores played an essential role in the formation process: First came the core, and then came the gas, attracted by the solid core. Thus all planet formation requires that a large lump of solid matter must form first. Of the Sun's planets, Jupiter has the largest of these cores, Saturn the next largest, Neptune the next, Uranus after that, and Earth ranks fifth, just as it does in total size. The formation histories of all the planets pose a fundamental question: How does nature make dust coagulate to form clumps of matter many thousand miles across?

The answer has two parts, one known and one unknown, with the unknown part, not surprisingly, closer to the origin. Once you form objects half a mile across, which astronomers call planetesimals, each of them will have sufficiently strong gravity to attract other such objects successfully. The mutual gravitational forces among planetesimals will build first planetary cores and then planets at a brisk pace, so that a few million years will take you from a host of clumps, each the size of a small town, to entire new worlds, ripe to acquire either a thin coat of atmospheric gases (in the case of Venus, Earth, and Mars) or an immensely thick one of hydrogen and helium (for the four gas-giant planets, which orbit the Sun at distances large enough for them to accumulate huge quantities of these two lightest gases). To astrophysicists, the tran-

sition from half-mile-wide planetesimals to planets reduces to a
series of well-understood computer models that produce a wide
variety of planetary details, but almost always yield inner planets
that are small, rocky, and dense, as well as outer planets that are
large and (except for their cores) gaseous and rarefied. During
this process, many of the planetesimals, as well as some of the
larger objects that they make, find themselves flung entirely out
of the solar system by gravitational interactions with still larger
objects.

All this works rather well on a computer, but building the half-
mile-wide planetesimals in the first place still lies beyond astro
astrophysicists' present abilities to integrate their knowledge of
physics with their computer programs. Gravity can't make planetesimals, because the modest gravitational forces between small
objects won't hold them together effectively. Two theoretical possibilities exist for making planetesimals from dust, neither of
them highly satisfactory. One model proposes the formation of
planetesimals through accretion, which occurs when dust particles collide and stick together. Accretion works well in principle,
because most dust particles *do* stick together when they meet.
This explains the origin of dust bunnies under your couch, and if
you imagine superdust bunnies growing around the Sun, you can,
with only minimal mental effort, let them grow to become chair-
sized, house-sized, block-sized, and before long the size of planetesimals, ready for serious gravitational action.

Unfortunately, unlike the production of actual bunnies, the
dust-bunny growth of planetesimals seems to require far too
much time. Radioactive dating of unstable nuclei detected in the
oldest meteorites implies that the formation of the solar system
required no more than a few tens of millions of years, and quite
possibly a good deal less time than that. In comparison with the
current age of the planets, approximately 4.55 billion years, this
amounts to a dram in the bucket, only 1 percent (or less) of the

total span of the solar system's existence. The accretion process requires significantly longer than a few tens of millions of years to make planetesimals from dust; so unless astrophysicists have missed something important in understanding how dust accumulates to build large structures, we need another mechanism to surmount the time barriers to planetesimal formation.

That other mechanism may consist of giant vortices that sweep up dust particles by the trillions, whirling them quickly toward their happy agglomeration into significantly larger objects. Because the contracting cloud of gas and dust that became the Sun and its planets apparently acquired some rotation, it soon changed its overall shape from spherical to platelike, leaving the Sun-in-formation as a relatively dense contracting sphere at the center, surrounded by a highly flattened disk of material in orbit around that sphere. To this day, the orbits of the Sun's planets, which all follow the same direction and lie in nearly the same plane, testify to a disklike distribution of the matter that built the planetesimals and planets. Within such a rotating disk, astrophysicists envision the appearance of rippling "instabilities," alternating regions of greater and lesser density. The denser parts of these instabilities collect both gaseous material and dust that floats within the gas. Within a few thousand years, these instabilities become swirling vortices that can sweep large amounts of dust into relatively small volumes.

This vortex model for the formation of planetesimals shows promise, though it has not yet won the hearts of those who seek explanations of how the solar system produced what young planets need. Upon detailed examination, the model provides better explanations for the cores of Jupiter and Saturn than for those of Uranus and Neptune. Because astronomers have no way to prove that the instabilities needed for the model to work actually did occur, we must refrain from passing judgment ourselves. The existence of numerous small asteroids and comets, which resemble planetesimals in their sizes and compositions, support the concept that billions of years ago, planetesimals by the millions built the planets. Let us therefore regard the formation of planetesimals as an established, if poorly understood, phenomenon that somehow bridges a key gap in our knowledge, leaving us ready to admire what happens when planetesimals collide.

In this scenario, we can easily imagine that once the gas and dust surrounding the Sun had formed a few trillion planetesimals, this armada of objects collided, built larger objects, and eventually created the Sun's four inner planets and the cores of its four giant planets. We should not overlook the planets' moons, smaller objects that orbit all of the Sun's planets except the innermost, Mercury and Venus. The largest of these moons, with diameters of a few hundred to a few thousand miles, appear to fit nicely into the model that we have created, because they presumably also arose from planetesimal collisions. Moon building ceased once collisions had built the satellite worlds to their present sizes, no doubt (we may assume) because by that time the nearby planets, with their stronger gravity, had taken possession of most of the nearby planetesimals. We should include in this picture the hundreds of thousands of asteroids that orbit between Mars and Jupiter. The largest of these, a few hundred miles in diameter, should likewise have grown through planetesimal collisions, and then have found themselves stymied from further growth by gravitational interference from the nearby giant planet Jupiter. The smallest asteroids, less than a mile across, may represent naked planetesimals, objects that grew from dust but never collided with one another, once again thanks to Jupiter's influence, after attaining sizes ripe for gravitational interaction.

For the moons that orbit the giant planets, this scenario seems to work quite well. All four giant planets have families of satellites that range in size from the large or extremely large (up to the size of Mercury) down to the small or even minuscule. The smallest of these moons, less than a mile across, may again be naked planetesimals, deprived of any further collisional growth by the presence of nearby objects that had already grown much larger. In each of these four families of satellites, almost all of the larger moons orbit the planet in the same direction and in nearly the same plane. We can hardly refrain from explaining this result with the same cause that made the planets orbit in the same direction and nearly the same plane: Around each planet, a rotating cloud of gas and dust produced clumps of matter, which grew to planetesimal and then to moon sizes.

In the inner solar system, only our Earth has a sizable moon. Mercury and Venus have none, while Mars' two potato-shaped moons, Phobos and Deimos, each span only a few miles, and should therefore represent the earliest stages of forming larger objects from planetesimals. Some theories assign the origin of these moons to the asteroid belt, with their present orbits around Mars the result of Mars' gravitational success in capturing these two former asteroids.

And what of our Moon, more than two thousand miles in diameter, surpassed in size only by Titan, Ganymede, Triton, and Callisto (and effectively tied with Io and Europa) among all the moons of the solar system? Did the Moon also grow from planetesimal collisions, as the four inner planets did?

This seemed quite a reasonable supposition until humans brought lunar rocks back to Earth for detailed examination. More than three decades ago, the chemical composition of the rock samples returned by the *Apollo* missions imposed two conclusions, one on either side of the possibilities for the Moon's origin. On the one hand, the composition of these Moon rocks resembles that of rocks on Earth so closely that the hypothesis that our satellite formed entirely apart from us no longer seems tenable. On the other hand, the Moon's composition differs from Earth's just enough to prove that the Moon did not entirely form from terrestrial material. But if the Moon did not form apart from Earth, and was not made from Earth, how did it form?

The current answer to this conundrum, amazing though it may seem on its surface, builds upon a once popular hypothesis that the Moon formed as the result of a giant impact, early in the history of the solar system, that scooped material out of the Pacific Basin and flung it into space, where it coalesced to form our satellite. Under the new view, which has already gained wide acceptance as the best available explanation, the Moon *did* form as the result of a giant object that struck Earth, but the object striking Earth was so large—about the size of Mars—that it naturally added some of its material to the matter ejected from Earth. Much of the material thrown into space by the force of the impact might have vanished from our immediate vicinity, but enough remained behind to coagulate into our familiar Moon, made of Earth plus foreign matter. All of this occurred 4.5 billion years ago, during the first 100 million years after the formation of the planets began.

If a Mars-sized object struck Earth in that bygone era, where is it today? The impact could hardly have knocked the object into pieces so small that we cannot observe them: our finest telescopes can find objects in the inner solar system as small as the planetesimals that built the planets. The answer to this objection takes us to a new picture of the early solar system, one that emphasizes its violent, collisional nature. The fact that planetesimals built a Mars-sized object, for example, did not guarantee that this object would endure for long. Not only did this object collide with Earth, but the good-sized pieces produced by that collision would also have continued to collide with Earth and the other inner planets, with each other, and with the Moon (once it had formed). In other words, collisional terror reigned over the inner solar system during its first several hundred million years, and the pieces of giant objects that struck the planets as they formed themselves became part of these planets. The Mars-sized object's impact on Earth merely ranked among the largest in a rain of bombardment, an epoch of destruction that brought planetesimals and much larger objects crashing down on Earth and its neighbors.

Seen from another perspective, this death-dealing bombardment simply marked the formation process's final stages. The process culminated in the solar system we see today, little changed during 4 billion years and more: one ordinary star, orbited by eight planets (plus icy Pluto, more akin to a giant comet than to a planet), hundreds of thousands of asteroids, trillions of meteoroids (smaller fragments that strike Earth by the thousands every day), and trillions of comets—dirty snowballs that formed at dozens of times Earth's distance from the Sun. We must not forget the planets' satellites, which have moved, with few exceptions, in orbits with long-term stability ever since their birth, 4.6 billion years ago. Let's take a closer look at the debris that continues to orbit our Sun, capable both of bringing forth life and of destroying life on the worlds such as ours.

CHAPTER 12

Between the Planets

From a distance, our solar system looks empty. If you enclosed it within a sphere large enough to contain the orbit of Neptune, then the Sun, together with all its planets and their moons, would occupy little more than one trillionth of all the space in that sphere. This result, however, assumes that interplanetary space is essentially empty. Viewed close up, however, the spaces between the planets turn out to contain all manner of chunky rocks, pebbles, ice balls, dust, streams of charged particles, and far-flung man-made probes. Interplanetary space is also permeated by immensely powerful gravitational and magnetic fields, invisible but nonetheless quite capable of affecting the objects in our neighborhood. These small objects and cosmic force fields present a serious ongoing threat to anyone who attempts to travel from place to place in the solar system. The largest of these objects likewise pose a threat to life on Earth, if they happen—as they certainly do on rare occasion—to collide with our planet at speeds of many miles per second.

Local regions of space are so not-empty that Earth, during its 30-kilometer-per-second orbital journey around the Sun, plows into hundreds of tons of interplanetary debris per day—most of it no larger than a grain of sand. Nearly all of this matter burns in Earth's upper atmosphere, slamming into the air with so much energy that the incoming particles vaporize. Our frail species evolved beneath this protective blanket of air. Larger, golf-ball-size pieces of debris heat rapidly but unevenly, and often shatter into many smaller pieces before they vaporize. Still larger pieces singe their surfaces but otherwise make their way, at least in part, down to the ground. You might think that by now, after 4.6 billion trips around the Sun, Earth would have "vacuumed" up all possible debris in its orbital path. We have made progress in this direction: things were once much worse. During the first half billion years after the formation of the Sun and its planets, so much junk rained down on Earth that the impact energy generated a strongly heated atmosphere and a sterilized surface.

In particular, one hunk of space junk was so substantial that it led to the formation of the Moon. The unexpected paucity of iron and other high-mass elements in the Moon, deduced from the lunar samples that the *Apollo* astronauts brought to Earth, indicates that the Moon most likely consists of matter spewn from Earth's relatively iron-poor crust and mantle by a glancing collision with a wayward, Mars-sized protoplanet. Some of the orbiting flotsam from this encounter coalesced to form our lovely, low-density satellite. Apart from this newsworthy event about 4.5 billion years ago, the period of heavy bombardment that Earth endured during its infancy was similar to that experienced by all the planets and other large objects in the solar system. They each sustained similar damage, with the airless, erosionless Moon and Mercury still preserving most of the craters produced during this period.

In addition to the flotsam left from its epoch of formation, interplanetary space also contains rocks of all sizes thrust from Mars, the Moon, and probably Earth as their surfaces reeled from high-energy impacts. Computer studies of meteor strikes demonstrate conclusively that some surface rocks near ground zero will be thrown upward with enough speed to escape the object's gravitational tether. From discoveries of Martian meteorites on Earth, we can conclude that about 1,000 tons of rocks from Mars rain down on Earth each year. Perhaps the same amount of debris reaches Earth from the Moon. Thus we did not have to go to the Moon to retrieve Moon rocks. A few dozen of them have come to us on Earth, although they are not of our choosing, and we had not yet learned this fact during the *Apollo* program.

If Mars ever harbored life—most likely billions of years ago when liquid water flowed freely on the Martian surface—then unsuspecting bacteria, stowed away in the nooks and crannies (especially in the crannies) of the rock ejected from Mars, could have traveled to Earth for free. We already know that some varieties of bacteria can survive long periods of hibernation, as well as high doses of the solar ionizing radiation to which they would be exposed en route to Earth. The existence of space-borne bacteria is neither a crazy idea nor pure science fiction. The concept even has an important-sounding name: panspermia. If Mars spawned life before Earth did, and if simple life traveled from Mars on ejected rocks and seeded Earth, we may all be descendants of Martians. This fact might seem to obviate environmental concerns over astronauts who sneeze on the Martian surface, spreading their germs on the alien landscape. In reality, even if we are all Martian in origin, we would dearly like to trace life's trajectory from Mars to Earth, so these concerns retain vital importance.

Most of the solar system's asteroids live and work in the "main belt," a flattened region around the Sun between the orbits of Mars and Jupiter. By tradition, asteroid discoverers get to name their objects as they choose. Often pictured by artists as a cluttered region of rocks floating in the plane of the solar system, though in fact spread out over millions of miles at different distances from the Sun, the objects in the asteroid belt have a total mass less than 5 percent of the Moon's, which itself has barely more than 1 percent of Earth's mass. Sounds insignificant at first, but the asteroids quietly pose a long-term cosmic threat to our planet. Accumulated perturbations of their orbits continually create a deadly subset of asteroids, perhaps a few thousand in number, whose elongated paths carry them so close to the Sun that they intersect the orbit of Earth, creating the possibility of collision. A back-of-the-envelope calculation demonstrates that most of these Earth-crossing asteroids will strike Earth within a few hundred million years. The objects larger than about a mile across carry enough energy to destabilize Earth's ecosystem and to put most of Earth's land species at risk of extinction. That would be bad.

Meanwhile, asteroids are not the only space objects that pose a risk to life on Earth. The Dutch astronomer Jan Oort first recognized that within the cold depths of interstellar space, much farther from the Sun than any planet, a host of frozen leftovers from the solar system's earliest stages of formation still orbit our star. This "Oort cloud" of trillions of comets extends to distances halfway to the closest stars, thousands of times larger than the size of the Sun's planetary system.

Oort's Dutch-American contemporary Gerard Kuiper proposed that some of these frozen objects once formed part of the disk of material from which the planets formed, and now orbit the Sun at distances considerably greater than Neptune's but much less than those of the comets in the Oort cloud. Collectively, they compose what astronomers now call the Kuiper Belt, a comet-strewn swath of circular real estate that begins just beyond the orbit of Neptune, includes Pluto, and extends several times as far again outward from Neptune as Neptune's distance from the Sun. The

most distant known object in the Kuiper Belt, named Sedna after
an Inuit goddess, has two-thirds of Pluto's diameter. Without a
nearby massive planet to perturb them, most of the Kuiper Belt
comets will maintain their orbits for billions of years. As in the
asteroid belt, a subset of the Kuiper Belt objects travel on eccen-
tric orbits that cross the paths of other planets. The orbit of Pluto,
which we may regard as an extremely large comet, as well as the
orbits of an ensemble of Pluto's small siblings, called Plutinos,
cross Neptune's path around the Sun. Other Kuiper Belt objects,
perturbed from their usual large orbits, occasionally plunge all
the way into the inner solar system, crossing planetary orbits with
abandon. This subset includes Halley, the most famous comet of
them all.

The Oort cloud is responsible for the long-period comets, those
whose orbital periods far exceed a human lifetime. Unlike Kuiper
Belt comets, Oort cloud comets can rain down on the inner solar
system from any angle and from any direction. The brightest
comet of the past three decades, comet Hyakutake (1996), came
from the Oort cloud, high above the plane of the solar system, and
will not return to our vicinity any time soon.

If we had eyes that could see magnetic fields, Jupiter would
look ten times larger than the full Moon in the sky. Spacecraft
that visit Jupiter must be designed to remain unaffected by this
powerful magnetism. As the English chemist and physicist
Michael Faraday discovered in 1831, if you move a wire across a
magnetic field, you will generate a voltage difference along the
wire's length. For this reason, fast-moving metal space probes can
have electrical currents induced within them. These currents
interact with the local magnetic field in a way that retards the
space probe's motion. This effect might explain the mysterious
slowing down of the two *Pioneer* spacecraft as they exit the solar
system. Both *Pioneer 10* and *Pioneer 11*, launched during the
1970s, have not traveled quite so far into space as our dynamical

models of their motions predict. After taking into account the effects of space dust encountered en route, along with recoils of the spacecraft arising from leaky fuel tanks, this concept of magnetic interaction—in this case with the Sun's magnetic field—may provide the best explanation for the slowdown of the *Pioneers*.

Better detection methods and close-flying space probes have increased the number of known planetary moons so rapidly that counting moons has become almost obsolete: they seem to multiply as we speak. What matters now is whether any of these moons are fun places to visit or to study. By some measures, the solar system's moons are far more fascinating than the planets they orbit. Mars' two moons, Phobos and Deimos, appear (not with those names) in Jonathan Swift's classic *Gulliver's Travels* (1726). Problem is, these two small moons were not discovered until more than a hundred years later; unless he was telepathic, Swift was presumably interpolating between Earth's single moon and Jupiter's (then known) four.

Earth's Moon has about 1/400 of the Sun's diameter, but is also just about 1/400 as far from us as the Sun, giving the Sun and the Moon the same size on the sky—a coincidence not shared by any other planet-moon combination in the solar system, and one that grants earthlings uniquely photogenic total solar eclipses. Earth has also locked onto the Moon's period of rotation, leaving the Moon's rotation period equal to its period of revolution around Earth. The capture has arisen from Earth's gravity, which exerts greater amounts of force on the denser parts of the Moon's interior and makes them always face toward Earth. Wherever and whenever this happens, as it does for Jupiter's four large moons, the locked moon shows only one face to its host planet.

Jupiter's system of moons stunned astronomers when they obtained their first good look. Io, the large moon closest to Jupiter, has been tidally locked and structurally stressed by its gravitational interactions with Jupiter and with the other large moons. These interactions pump enough energy into Io (about the same size as our Moon) to melt some of its rocky interior, making Io the most volcanically active object in the solar system. Jupiter's second large moon, Europa, has enough H_2O that its internal heat, which arises from the same interactions that affect Io, has melted its subsurface ice, leaving a liquid ocean below an icy covering.

Close-up images of the surface of Miranda, one of Uranus' moons, reveal badly mismatched patterns, as though the poor moon had been blown apart, and its pieces hastily glued back together. The origin of these exotic features remains a mystery, but may also be due to something simple, like the uneven upwelling of ice sheets.

Pluto's lone moon, Charon, is so large and so close to Pluto that Pluto and Charon have tidally locked onto each other—both objects have rotation periods equal to their periods of revolution around their common center of mass. By convention, astronomers name planets' moons after significant Greek personalities in the life of the god whose name the planet bears, though they use the Roman counterpart's name for the planet itself (Jupiter rather than Zeus, for example). Because the classical gods led complicated social lives, no shortage of characters exists from which to draw names.

Sir William Herschel was the first person to discover a planet beyond those easily visible to the naked eye, and he was ready to name this new planet after the king who might support his research. Had Sir William succeeded, the planet list would read: Mercury, Venus, Earth, Mars, Jupiter, Saturn, and George. Fortunately, clearer heads prevailed, so that some years later the new planet received the classical name Uranus. But Herschel's original suggestion to name the planet's moons after characters in William Shakespeare's plays and Alexander Pope's poem *The Rape of the Lock* remains the tradition to this day. Among Uranus' seventeen moons we find Ariel, Cordelia, Desdemona, Juliet, Ophelia, Portia, Puck, and Umbriel, with two new moons, Caliban and Sycorax, discovered as recently as 1997.

The Sun loses material from its surface at a rate of 200 million tons per second (which happens to closely match the rate at which water flows through the Amazon Basin). The Sun loses this mass in the "solar wind," which consists of high-energy charged particles. Traveling up to 1,000 miles per second, these particles stream through interplanetary space, where they are often deflected by planetary magnetic fields. In response, these particles spiral down toward a planet's north and south magnetic poles, colliding with atmospheric gas molecules to produce colorful auroral glows. The Hubble Space Telescope has spotted aurorae near the poles of both Saturn and Jupiter. On Earth, the aurorae borealis and australis (the Northern and Southern lights) serve as intermittent reminders of how sweet it is to have a protective atmosphere.

Earth's atmosphere technically extends much farther above Earth's surface than we generally conceive. Satellites in "low-Earth orbit" typically travel at altitudes of 100 to 400 miles and complete an orbit in about 90 minutes. Although no one can breathe at these altitudes, some atmospheric molecules remain— enough to drain orbital energy slowly from unsuspecting satellites. To combat this drag, satellites in low orbit require intermittent boosts, lest they fall back to Earth and burn up in the atmosphere. The most sensible way to define the edge of our atmosphere is to ask where the density of its gas molecules falls to the density of gas molecules in interplanetary space. With this definition, Earth's atmosphere extends thousands of miles into space. Orbiting high above this level, 23,000 miles above Earth's surface (one tenth of the distance to the Moon), are the communications satellites that carry news and views around Earth. At

this special altitude, a satellite finds not only that Earth's atmosphere is irrelevant but also that its speed in orbit, thanks to the
diminished pull from Earth at this greater distance from our
planet, falls to the point that it takes twenty-four hours to complete each revolution around our planet. Moving in orbits that
precisely match Earth's rotation rate, these satellites appear to
"hover" above a single point on the Equator, a fact that makes
them ideal for relaying signals from one part of Earth's surface to
another.

Newton's law of gravity states that, although the gravity from
a planet gets progressively weaker as you travel farther from it, no
distance will reduce the force of gravity all the way to zero, and
that an object with enormous mass can exert significant gravitational forces even at large distances. The planet Jupiter, with its
mighty gravitational field, bats out of harm's way many comets
that would otherwise wreak havoc on the inner solar system. By
doing so, Jupiter acts as a gravitational shield for Earth, allowing
long (50- to 100-million-year) stretches of relative peace and
quiet on Earth. Without Jupiter's protection, complex life would
have a hard time growing interestingly complex, always living at
the risk of extinction from a devastating impact.

We have exploited the gravitational fields of planets for nearly
every probe sent into space. The *Cassini* probe, for example, sent
to Saturn for an encounter late in 2004, was launched from Earth
on October 15, 1997, and was gravitationally assisted twice by
Venus, once by Earth (on a return flyby), and once by Jupiter. Like
a multi-cushion billiard shot, trajectories from one planet to
another using gravitational slingshots are common. Otherwise
our tiny probes would not have enough speed and energy to reach
their destinations.

One of us is now accountable for a piece of the solar system's
interplanetary debris. In November 2000, the main-belt asteroid
1994KA, discovered by David Levy and Caroline Shoemaker, was

named "13123 Tyson." A fun distinction, but there's no particular reason to get big-headed about it; as already noted, plenty of asteroids have familiar names such as Jody, Harriet, and Thomas. And plenty of other asteroids have names such as Merlin, James Bond, and Santa. Rising through 20,000, the count of asteroids with well-established orbits (the criterion for assigning them names and numbers) may soon challenge our capacity to name them. Whether or not that day arrives, there is curious comfort knowing that one's own chunk of cosmic debris is not alone, as it litters the space between the planets, joined by a long list of other chunks named for real and fictional people.

When last checked, asteroid 13123 Tyson was not headed toward us, and so cannot be blamed for either ending or starting life on Earth.

CHAPTER 13

Worlds Unnumbered

Planets Beyond the Solar System

Thro' worlds unnumbered tho' the God be known,
'Tis ours to trace him only in our own.
He, who through vast immensity can pierce,
See worlds on worlds compose one universe,
Observe how system into system runs
What other planets circle other suns,
What varied Being peoples ev'ry star,
May tell why Heav'n has made us as we are.

—Alexander Pope, *An Essay on Man* (1733)

Nearly five centuries ago, Nicolaus Copernicus resurrected a hypothesis that the ancient Greek astronomer Aristarchus had first suggested. Far from occupying the center of the cosmos, said Copernicus, Earth belongs to the family of planets that orbit the Sun.

Even though a majority of humans have yet to accept this fact, believing in their hearts that Earth remains immobile as the heavens turn around her, astronomers have long offered convincing arguments that Copernicus wrote the truth about the nature of our cosmic home. The conclusion that Earth ranks as just one of the Sun's planets immediately suggests that other planets fundamentally resemble our own, and that they may well possess their own inhabitants, endowed as we are with plans and dreams, work, play, and fantasy.

For many centuries, astronomers who used telescopes to observe hundreds of thousands of individual stars lacked the ability to discern whether or not any of these stars have planets of their own. Their observations did reveal that our Sun ranks as an entirely representative star, whose near twins exist in great numbers throughout our Milky Way galaxy. If the Sun has a planetary family, so too might other stars, with their planets equally capable of giving life to creatures of all possible forms. Expressing this view in a manner that affronted papal authority brought Giordano Bruno to his death at the stake in 1600. Today, a tourist can pick his way through the crowds at the outdoor cafés in Rome's Campo di Fiori to reach Bruno's statue at its center, then pause for a moment to reflect on the power of ideas (if not the power of those who hold them) to triumph over those who would suppress them.

As Bruno's fate helps to illustrate, imagining life on other worlds ranks among the most powerful ideas ever to enter human minds. Were this not so, Bruno would have lived to a riper age, and NASA would find itself shorter of funds. Thus speculation about life on other worlds has focused throughout history, as NASA's attention still does, on the planets that orbit the Sun. In our search for life beyond Earth, however, a great frost has appeared: none of the other worlds in our solar system seem particularly fit for life.

Although this conclusion hardly does justice to the myriad possible paths by which life might arise and maintain itself, the fact remains that our initial explorations of Mars and Venus, as well as of Jupiter and its large moons, have failed to produce any convincing signs of life. To the contrary, we have found a great deal of evidence for conditions extremely hostile to life as we know it. Much more searching remains to be done, and fortunately (for those who engage themselves mentally in this effort) continues to be underway, especially in the hunt for life on Mars. Nevertheless, the verdict on extraterrestrial life in the solar system shows enough likelihood of proving negative that supple minds now usually look beyond our cosmic neighborhood, to the vast array of possible worlds that orbit stars other than our Sun.

Until 1995, speculation about planets around other stars could
proceed almost entirely unfettered by facts. With the exception of
a few pieces of Earth-sized debris in orbit around the remnants of
exploded stars, which almost certainly formed after the supernova
explosion and barely qualified as planets, astrophysicists had
never found a single "exosolar planet," a world orbiting a star
other than the Sun. At the end of that year came the dramatic
announcement of the first such discovery; then, a few months
later, came four more; and then, with the floodgates open, finding
new worlds proceeded ever more swiftly. Today, we know of far
more exosolar planets around other stars than of the now familiar worlds that orbit the Sun—a tally that exceeds 100 and is
almost certain to keep growing for years to come.

To describe these newfound worlds, and analyze the implications of their existence in the search for extraterrestrial life, we
must confront a single hard-to-believe fact: Although astrophysicists assert that they not only know that these planets exist but
have also deduced their masses, their distances from their parent
stars, the times that the planets take to complete their orbits, and
even the shapes of those orbits, no one has ever seen or photographed a single one of these exosolar planets.

How can anybody deduce so much about planets they have
never seen? The answer lies in detective work familiar to those
who study starlight. By separating that light into its spectrum of
colors, and by comparing those spectra among thousands of stars,
those who specialize in observing starlight can recognize different
types of stars purely by the ratios of the intensities of the different colors that appear in stellar spectra. Once upon a time, these
astrophysicists photographed the stars' spectra, but today they use
sensitive devices that register digitally how much starlight of
each particular color reaches us on Earth. Though the stars are

many trillions of miles from us, their fundamental natures have become an open book. Astrophysicists can now easily determine—purely by measuring the spectrum of the colors of starlight—which stars most closely resemble the Sun, which are somewhat hotter and more luminous, and which are cooler and intrinsically fainter than our star.

But they can also do more. Having grown familiar with the distribution of colors in the spectra of various types of stars, astrophysicists can quickly identify a familiar pattern in the star's spectrum, which typically shows the partial or total absence of light at particular colors. They often recognize such a pattern, but find that all the colors that form it have been slightly shifted toward either the red or the violet end of the spectrum, so that all the familiar guideposts are now either somewhat redder or somewhat more violet than the norm.

Scientists characterize these colors by their wavelengths, which measure the separation between successive wave crests in the vibrating light waves. Because they correspond with the colors that our eyes and brains perceive, specifying exact wavelengths simply names colors more precisely than we do in normal speech. When astrophysicists spot a familiar pattern in the intensity of light measured for thousands of different colors, but find that all the wavelengths in the pattern are (for example) 1 percent longer than usual, they conclude that the star's colors have changed as the result of the Doppler effect, which describes what happens when we observe an object either approaching us or receding from us. If, for example, an object moves toward us, or we move toward it, we find that all the wavelengths of the light that we detect are *shorter* than those we measure from an identical object at rest with respect to ourselves. If the object recedes from us, or we recede from it, we find all the wavelengths to be *longer* than those from an object at rest. The deviation from the at-rest situation depends on the relative velocity between the light source and those who observe it. For speeds much less than the speed of light (186,000 miles per second), the fractional change in all the wavelengths of light, called the Doppler shift, equals the ratio of the speed of approach or recession to the speed of light.

During the 1990s, two teams of astronomers, one in the United States and one centered in Switzerland, devoted themselves to increasing the precision with which they could measure the Doppler shifts of starlight. They did so not simply because scientists always prefer to make more accurate measurements, but because they had a straightforward goal: to detect the existence of *planets* by studying the light from *stars*.

Why this roundabout approach to the detection of exosolar planets? Because for now this method offers the only effective way to discover them. If our solar system offers any guide to the distances at which planets orbit stars, we must conclude that these distances amount to only a tiny fraction of the distances between stars. The Sun's closest neighbor stars are about half a million times farther from us than the distance between the Sun and its innermost planet, Mercury. Even Pluto's distance from the Sun is less than one five-thousandth of the distance to Alpha Centauri, our closest star system. These astronomically minuscule separations between the stars and their planets, combined with the faintness with which a planet reflects light from its star, make it nearly impossible for us to actually see any planets beyond the solar system. Imagine, for example, an astrophysicist on a planet around one of the Alpha Centauri stars who turns her telescope toward the Sun and attempts to spot Jupiter, the Sun's largest planet. The Sun-Jupiter distance amounts to only one fifty-thousandth of the distance to the Sun, and Jupiter shines with just one billionth of the Sun's intensity. Astrophysicists like to compare this to the problem of seeing a firefly next to a searchlight's glare. We may do it some day, but for now the quest to observe exosolar planets directly lies beyond our technological capabilities.

The Doppler effect offers another approach. If we study the star closely, we can carefully measure any changes that appear in the Doppler shift of the light from that star. These changes must arise from changes in the speed with which the star is either approaching us or receding from us. If the changes prove to be cyclical—that is, if their amounts rise to a maximum, fall to a minimum, rise to the same maximum again, and repeat this cycle over the same intervals of time—then the entirely reasonable conclusion follows that the star must be moving in an orbit that takes it around and around some point in space.

What could make a star dance like that? Only the gravitational force from another object, so far as we know. No doubt that planets, by definition, have masses much less than the mass of a star, so they exert only modest amounts of gravitational force. When they pull on a nearby star that possesses far more mass than they do, they produce only small changes in the star's velocity. Jupiter, for example, changes the Sun's velocity by about 40 feet per second, slightly more than the speed of a world-class sprinter. As Jupiter performs its twelve-year orbit around the Sun, an observer located along the plane of this orbit would measure Doppler shifts in the Sun's light. These Doppler shifts would demonstrate that at a particular time, the Sun's velocity with respect to the observer would rise 40 feet per second above its average value. Six years later, the same observer would find that the Sun's velocity is 40 feet per second less than average. During the interim, this relative velocity would shift smoothly between its two extreme values. After a few decades of observing this repetitive cycle, the observer would justifiably conclude that the Sun has a planet moving in a twelve-year orbit that causes the Sun to perform its own orbit, producing the velocity changes that arise naturally from this motion. The size of the Sun's orbit, in comparison to the size of Jupiter's, exactly equals the *inverse* of the ratio of the two objects' masses. Since the Sun has one thousand times Jupiter's mass,

Jupiter's orbit around their mutual center of gravity is one thousand times *larger* than the Sun's—testimony to the fact that the Sun is a thousand times more difficult to budge than Jupiter.

Of course, the Sun has many planets, each of which simultaneously pulls on the Sun with its own gravitational force. The Sun's net motion therefore amounts to a superposition of orbital dances, each with a different cyclical period of repetition. Because Jupiter, the Sun's largest and most massive planet, exerts the greatest amount of gravitational force on the Sun, the dance imposed by Jupiter dominates this complex pattern.

When astrophysicists sought to detect exosolar planets by watching stars dance, they knew that to find a planet roughly similar to Jupiter, orbiting its star at a distance comparable to Jupiter's distance from the Sun, they would have to measure Doppler shifts with an accuracy sufficient to reveal velocity changes of approximately 40 feet per second. On Earth this sounds like a significant speed (about 27 miles per hour), but in astronomical terms, we are talking about less than one millionth of the speed of light, and about one thousandth of the typical speed with which stars happen to be moving toward us or away from us. Thus to detect the Doppler shift produced by a change in velocity equal to one millionth of the speed of light, astrophysicists must measure changes in wavelength—that is, in star colors—of one part in a million.

These precision measurements yielded more than the detection of planets. First of all, because the detection scheme lies in finding a cyclical repetition in the changes of a star's velocity, the length of each of these cycles directly measures the orbital period of the planet responsible for it. If the star dances with a particular cycle of repetition, the planet likewise must be dancing with an identical period of motion, though in a much larger orbit. This orbital period in turn reveals the distance of the planet from the star. Isaac Newton long ago proved that an object orbiting a star will complete each orbit more rapidly when closer to the star, more slowly when farther away: each orbital period corresponds to a particular value of the average distance between the star and the orbiting object. In the solar system, for instance, a one-year orbital period implies a distance equal to the Earth-Sun distance, whereas a twelve-year period implies a distance 5.2 times larger, the size of Jupiter's orbit. So the research team could announce not only that they had found a planet but also that they knew both the planet's orbital period and its average distance from its star.

They could deduce still more about the planet. Moving at a particular distance from its star, a planet's gravity will pull on the star with a force that depends on the planet's mass. More massive planets exert greater force, and these forces make the star dance more rapidly. Once they knew the planet-star distances, the team could then include the *masses* of the planets in the list of planetary characteristics that they had determined through careful observation and deduction.

This deduction of a planet's mass by observing the star's dance comes with a disclaimer. Astronomers have no way to tell whether they are studying a dancing star from a direction that happens to coincide exactly with the plane of the planet's orbit, or from a direction directly above the plane of the orbit (in which case they will measure a zero velocity for the star), or (in almost all cases) from a direction neither exactly along the plane nor directly perpendicular to it. The plane of the planet's orbit around the star coincides with the plane of the star's motion in response to the planet's gravity. We therefore observe the full orbital speeds only if our line of sight to the star happens to be the same as the plane of the planet's orbit around the star. To imagine a loosely analogous situation, put yourself at a baseball game, able to measure the speed of the pitched ball as it comes toward you or moves away, but not the speed with which the ball crosses your field of vision. If you are a talent scout, the best place for you to sit is behind home plate, in direct line with the baseball's motion. But if you observe the game from the first or third baselines, the ball thrown by the pitcher will, for the most part, neither approach you nor recede from you, so your measurement of the ball's speed along your line of sight will be nearly zero.

Because the Doppler effect reveals only the speed with which a star moves toward us or away from us, but not how rapidly the star crosses our line of sight, we usually cannot tell how nearly our line of sight to the star lies in the plane of the star's orbit. This fact implies that the masses that we deduce for exosolar planets are all *minimum* masses; they will prove to be the planets' actual masses only in those cases when we do observe the star along its orbital plane. On the average, the actual mass of an exosolar planet equals twice the minimum mass deduced from observing the star's motions, but we have no way to know which exosolar planet masses lie above this average ratio, and which below.

In addition to deducing the planet's orbital period and orbital size, as well as the planet's minimum mass, astrophysicists who study star dances by the Doppler effect have one more success: they can determine the shape of the planet's orbit. Some of these orbits, like those of Venus and Neptune around the Sun, have an almost perfect circularity; but others, like the orbits of Mercury, Mars, and Pluto, have significant elongation, with the planet traveling much closer to the Sun at some points along its orbit than at others. Because a planet moves more rapidly when it is closer to its star, the star changes its velocity more rapidly at those times. If astronomers observe a star that changes its velocity at a constant rate throughout its cyclical period, they conclude that these changes arise from a planet moving in a circular orbit. If, on the other hand, they find that the changes sometimes occur more rapidly and sometimes more slowly, they deduce that the planet has a noncircular orbit, and can find the amount of the orbital elongation—the amount by which the orbit deviates from circularity—by measuring the different rates at which the star changes its velocity throughout the orbital cycle.

Thus, in a triumph of accurate observations coupled with their powers of deduction, astrophysicists who study exosolar planets can provide four key properties of any planet that they find: the planet's orbital period; its average distance from its star; its minimum mass; and its orbital elongation. Astrophysicists achieve all this by capturing the colors of light from stars that lie hundreds of trillions of miles from the solar system, and by measuring those changes with a precision better than one part in a million—a high point in our attempts to probe the heavens in a search for Earth's cousins.

Only one problem remains. Many of the exosolar planets discovered during the past decade orbit their stars at distances much smaller than any of the distances between the Sun to its planets. This issue looms larger because all the exosolar planets so far detected have masses comparable to that of Jupiter, a giant planet that orbits the Sun at more than five times the Earth-Sun distance. Let us take a moment to examine the facts, before we admire the astrophysicists' explanations of how these planets may have come to occupy orbits so much smaller than those familiar to us in our own planetary system.

Whenever we use the star dance method to search for planets around other stars, we must remain aware of the biases built into this method. First, planets close to their stars take much less time to orbit than do planets far from their stars. Since astrophysicists have limited amounts of time with which to observe the universe, they will naturally discover planets moving in, for example, six-month periods far more quickly than they can detect planets that take a dozen years for each orbit. In both cases, the astrophysicists must wait through at least a couple of orbits to be certain that they have detected a repeatable pattern of the changes in the stars' velocities. To find planets with orbital periods comparable to Jupiter's twelve years could therefore consume much of an individual's professional career.

Second, a planet will exert more gravitational force on its host star when close rather than when far. These greater forces make the star dance more rapidly, producing larger Doppler shifts in their spectra. Since we can detect larger shifts more easily than smaller ones, the closer-in planets attract more attention, and do so more rapidly, than the farther-out planets do. At all distances, however, an exosolar planet must have a mass roughly comparable to Jupiter's (318 times Earth's) to be detected by the Doppler shift method. Planets with significantly less mass cannot make their stars dance with a speed that rises above the threshold of detectability by today's technology.

In hindsight, then, no surprise should have accompanied the news that the first exosolar planets to be discovered all have masses comparable to Jupiter's, and all orbit close to their stars. The surprise lay in just how close may of these planets turned out to be—so close that they take not several months or years to complete each orbit, as the Sun's planets do, but only a few days. Astrophysicists have now found more than a dozen planets that complete each orbit in less than a week, with the record holder sweeping out each orbit in just over two and a half days. This planet, which orbits the Sun-like star known as HD73256, has a mass at least 1.9 times Jupiter's mass, and moves in a slightly elongated orbit at an average distance from its star equal to only 3.7 percent of the Earth-Sun distance. In other words, this giant planet possesses more than 600 times Earth's mass at a distance from its star less than one tenth that of Mercury.

Mercury consists of rock and metal, baked to temperatures of many hundred degrees on the side that happens to face the Sun. In contrast, Jupiter and the Sun's other giant planets (Saturn,

Uranus, and Neptune) are enormous balls of gas, surrounding solid cores that include only a few percent of each planet's mass. All theories of planet formation imply that a planet with a mass comparable to Jupiter's cannot be solid, like Mercury, Venus, and Earth, because the primordial cloud that formed planets contained too little of the stuff that can solidify to make a planet with more than a few dozen times the mass of Earth. The conclusion follows, as one more step in the great detective story that has given us exosolar planets, that all exosolar planets so far discovered (since they have masses comparable to Jupiter's), must likewise be great balls of gas.

Two questions immediately arise from this startling conclusion: How did these Jupiter-like planets ever come to orbit so close to their stars, and why doesn't their gas quickly evaporate under the intense heat? The second question has a relatively easy answer: The planets' enormous masses can retain even light gases heated to temperatures of hundreds of degrees, simply because the planets' gravitational forces can overcome the tendency of the atoms and molecules in the gas to escape into space. In the most extreme cases, however, this contest tips only narrowly in favor of gravity, and the planets lie just outside the distance at which their stars' heat would indeed evaporate their gases.

The first question, of how giant planets came to orbit so close to Sun-like stars, brings us to the fundamental issue of how planets formed. As we have seen in Chapter 11, theorists have worked hard to achieve some understanding of the planet-formation process in our solar system. They conclude that the Sun's planets accumulated themselves into being, growing from smaller clumps of matter into larger ones within a pancake-shaped cloud of gas and dust. Within this flattened, rotating mass of matter that surrounded the Sun, individual concentrations of matter formed, first at random, but then, because they had a density greater than average, by winning the gravitational tug-of-war among particles.

In the final stages of this process, Earth and the other solid planets survived an intense bombardment from the last of the giant chunks of material.

As this agglutinative process unfolded, the Sun began to shine, evaporating the lightest elements, such as hydrogen and helium, from its immediate neighborhood, and leaving its four inner planets (Mercury, Venus, Earth, and Mars) composed almost entirely of heavier elements such as carbon, oxygen, silicon, aluminum, and iron. In contrast, each of the clumps of matter that formed at five to thirty times Earth's distance from the Sun remained sufficiently cool to retain much of the hydrogen and helium in its vicinity. Because these two lightest elements are also the most abundant, this retentive ability produced four giant planets, each with many times Earth's mass.

Pluto belongs neither to the class of rocky, inner planets nor to the group of outer gas-giant planets. Instead, Pluto, still uninspected by spacecraft from Earth, resembles a giant comet, made of a mixture of rock and ice. Comets, which typically have diameters of 5 to 50 miles rather than Pluto's 2,000 miles, rank among the first sizable clumps of matter to form within the early solar system; they are rivaled in age by the oldest meteorites, which are fragments of rock, metal, or rock-and-metal mixtures that happen to have struck Earth's surface and to have been recognized by those who know how to tell a meteorite from a garden-variety rock.

Thus the planets built themselves from matter much like that in comets and meteorites, with the giant planets using their solid cores to attract and retain a much larger amount of gas. Radioactive dating of the minerals in meteorites have shown that the oldest of them have ages of 4.55 billion years, significantly greater than the oldest rocks found on the Moon (4.2 billion years) or on Earth (just less than 4 billion years). The birth of the solar system, which therefore occurred about 4.55 billion B.C., quite naturally led to the segregation of planetary worlds into two groups:

the relatively small, solid inner planets and the much larger, more massive, mainly gaseous giant planets. The four inner planets orbit the Sun at distances of 0.37 to 1.52 times the Earth-Sun distance, while the four giants remain at the much greater distances, ranging from 5.2 to 30 times the Earth-Sun distance, which allowed them to be giants.

This description of how the Sun's planets formed makes such good sense that it almost seems a shame that we have now found so many examples of objects with masses similar to Jupiter's, moving in orbit around their stars at distances much less than Mercury's distance from the Sun. Indeed, because the first exosolar planets to be discovered all had such small distances from their stars, for a time it appeared as though our solar system might prove the exception, rather than the model of planetary systems, as theorists had implicitly assumed in the days when they had nothing else on which to base their conclusions. Understanding the bias imposed by the relative ease of discovering close-in planets gave them some reassurance, and before long they had observed for sufficiently long times, and with sufficient accuracy, to detect gas-giant planets at much greater distances from their stars.

Today, the list of exosolar planets, ordered by distance from the star to the planet, begins with the entry described above, of a planet that takes only 2.5 days to perform each orbit, and extends, through well over a hundred entries, to the star 55 Cancri, where a planet with at least four times the mass of Jupiter takes 13.7 years for each orbit. Astrophysicists can calculate from the orbital period that this planet has a distance from its star equal to 5.9 times the Sun-Earth distance, or 1.14 times the distance from the Sun to Jupiter. The planet ranks as the first to be found with a distance from its star greater than the Sun-Jupiter distance, and therefore seems to provide a planetary system roughly comparable to our solar system, at least so far as the star and its largest planet are concerned.

However, this is not quite so. The planet that orbits 55 Cancri at 5.9 times the Earth-Sun distance represents not the first but the *third* to be discovered in orbit around this star. By now, astronomers have accumulated sufficient data, and have grown so skilled at interpreting their Doppler shift observations, that they can disentangle the complex star dance produced by two or more planets. Each of these planets attempts to impose a dance in its own rhythm, with a repetitive period equal to the span of the planet's orbit around the star. By observing for a sufficiently long time, and by employing computer programs that fear no calculation, planet hunters can tease from comingled dances the basic steps induced by each orbiting world. In the case of 55 Cancri, a modest star visible in the constellation called the Crab, they had already found two closer-in planets, with orbital periods of 42 days and 89 days and minimum masses of 0.84 and 0.21 Jupiter masses, respectively. The planet with a minimum mass equal to "only" 0.21 Jupiter masses (67 Earth masses) ranks among the least massive yet detected; but the record low mass for an exosolar planet has now fallen to 35 Earth masses—still so many times greater than Earth's that we should not hold our breath in anticipation that astronomers will soon find Earth's cousins.

Circle it as we may, we cannot avoid the problem, evident from the orbits of the planets around 55 Cancri, of explaining why and how many exosolar planets, with masses much like Jupiter's, orbit their stars at stunningly small distances. No planet with a Jupiter-like mass can form, experts will tell you, much closer to a Sun-like star than three to four times the Earth-Sun distance. If we assume that exosolar planets obey this dictum, they must have somehow moved to much smaller distances after they had formed. This conclusion, if valid, raises at least three burning questions:

1. What made these planets move into smaller orbits after they had formed?

2. What stopped them from moving all the way into their stars and perishing?

3. Why did this occur in many other planetary systems, but not in our solar system?

These questions have answers, supplied by fertile minds once they had been properly stimulated by the discovery of exosolar planets. We may summarize the scenario now favored by experts as follows:

1. "Planetary migration" occurred because significant amounts of material left over from the formation process continued to orbit the star within the orbits of the new-formed giant planets. This material gets systematically flung by the big planet's gravity to outer orbits, which in turn forces the big planet to creep inward.

2. When the planets had approached much closer to their stars than their points of origin, the tidal forces from the star locked the planet into place. These forces, comparable to the tidal forces from the Sun and Moon that raise tides in Earth's oceans, forced the planets' rotational periods to equal their orbital periods, as happened to the Moon from Earth's tidal forces. They also prevented any further approach of the planet to the star, for reasons that require sufficient involvement with celestial mechanics to merit passing over here.

3. Presumably the luck of the draw determined which planetary systems formed with large amounts of debris, capable of inducing planetary migration, and which, like our own, had relatively little debris, so the planets remained at the distances at which they had formed. In the case of the planets around 55 Cancri, it is possible that all three migrated significantly inward, with the outermost planet having formed at several times its current distance from the star. Or it may be that the details of how much debris lived inside the planet's orbit, and how much outside,

caused significant migration of the two inner planets, while the third has remained in its original path.

Some work remains to be done, to put things politely, before astrophysicists can proclaim they have explained how planetary systems form around stars. Meanwhile, those who hunt for exosolar planets continue to pursue their dream of finding Earth's twin, a planet similar to Earth in its size, mass, and orbital distance from its parent star. When and if they find such a planet, they hope to examine it—even from a distance of dozens of light-years—with sufficient precision to determine whether the planet possesses an atmosphere and oceans similar to Earth's, and perhaps whether life exists upon that planet like our own.

In pursuit of this dream, astrophysicists know that they need instruments orbiting above our atmosphere, whose blurring effects prevent us from making extremely precise measurements. One experiment, NASA's *Kepler* mission, aims at observing hundreds of thousands of nearby stars, seeking the tiny diminution in starlight (about one hundredth of 1 percent) caused by the motion of an Earthsized planet across our line of sight to a star. This approach can succeed only for the small fraction of situations in which our view lies almost exactly along the planet's orbital plane, but for those cases, the interval between planetary transits equals the planet's orbital period, which in turn specifies the planet-star distance, and the amount of starlight diminution reveals the size of the planet.

However, if we hope to find out more than the planet's bare physical characteristics, we must study the planet by direct imaging and analysis of the spectrum of the light that the planet reflects into space. NASA and ESA, the European Space Agency, have programs under way to achieve this goal within two decades. To see another Earth-like planet, even as a pale blue dot close to a far brighter star, could inspire another generation of poets, physicists, and politicians. To analyze the planet's reflected light, and thus to determine whether or not the planet's atmosphere contains oxygen (a likely indication of life) or oxygen plus methane (an almost completely definitive mark of life), would mark the sort of accomplishment that the bards once sang, elevating mere mortals into heroes for the ages, leaving us face to face (as F. Scott Fitzgerald wrote in *The Great Gatsby*) with something commensurate with man's capacity to wonder. For those who dream of finding life elsewhere in the universe, our final section awaits.

Part V

The Origin
of Life

CHAPTER 14

Life in the Universe

Our survey of origins brings us, as we knew it would, to the most intimate and arguably the greatest mystery of all: the origin of life, and in particular of forms of life with which we may someday communicate. For centuries, humans have wondered how we might find other intelligent beings in the cosmos, and with whom we might enjoy at least a modest conversation before we pass into history. The crucial clues for resolving this puzzle may appear in the cosmic blueprint of our own beginnings, which includes Earth's origin within the Sun's family of planets, the origin of the stars that provide energy for life, the origin of structure in the universe, and the origin and evolution of the universe itself.

If we could only read this blueprint in detail, it could direct us from the largest to the smallest astronomical situations, from the unbounded cosmos to individual locations where different types of life flourish and evolve. If we could compare the diverse forms of life that arose under various circumstances, we could perceive

the rules of life's beginnings, both in general terms and in particular cosmic situations. Today, we know of only one form of life:
life on Earth, all of which shares a common origin and uses DNA
molecules as the fundamental means of reproducing itself. This
fact deprives us of multiple examples of life, relegating to the
future a general survey of life in the cosmos, unachievable until
the day we begin to discover forms of life beyond our planet.

Things could be worse. We do know a great deal about life's history on our planet, and must build on this knowledge to derive
basic principles about life throughout the universe. To the extent
that we can rely on these principles, they will tell us when and
where the universe provides, or has provided, the basic requirements for life. In all our attempts to imagine life elsewhere, we
must resist falling into the trap of anthropomorphic thinking, our
natural tendency to imagine that extraterrestrial forms of life
must be much like our own. This entirely human attitude, which
arises from our evolutionary and personal experiences here on
Earth, restricts our imagination when we attempt to conceive how
different life on other worlds may be. Only biologists familiar
with the amazing variety and appearance of different forms of
life on Earth can confidently extrapolate what extraterrestrial
creatures might look like. Their strangeness almost certainly lies
beyond the imaginative powers of ordinary humans.

Some day—perhaps next year, perhaps during the coming century, perhaps long after that—we shall either discover life beyond
Earth or acquire sufficient data to conclude, as some scientists
now suggest, that life on our planet represents a unique phenomenon within our Milky Way galaxy. For now, our lack of
information on this subject allows us to consider an enormously
broad range of possibilities: We may find life on several objects in
the solar system, which would imply that life probably exists
within billions of similar planetary systems in our galaxy. Or we
may find that Earth alone has life within our solar system, leav-

ing the question of life around other stars open for the time being. Or we may eventually discover that life exists nowhere around other stars, no matter how far and wide we look. In the search for life in the universe, just as in other spheres of activity, optimism feeds on positive results, while pessimistic views grow stronger from negative outcomes. The most recent information that bears upon the chances for life beyond Earth—the discovery that planets are moving in orbit around many of the Sun's neighboring stars—points toward the optimistic conclusion that life may prove relatively abundant in the Milky Way. Nevertheless, great issues remain to be resolved before this conclusion can gain a firmer footing. If, for example, planets are indeed abundant, but almost none of these planets provide the proper conditions for life, then the pessimistic view of extraterrestrial life seems likely to prove correct.

Scientists who contemplate the possibilities of extraterrestrial life often invoke the Drake equation, after Frank Drake, the American astronomer who created it during the early 1960s. The Drake equation provides a useful concept rather than a rigorous statement of how the physical universe works. The equation usefully organizes our knowledge and ignorance by separating the number that we dearly seek to estimate—the number of places where intelligent life now exists in our galaxy—into a set of terms, each of which describes a necessary condition for intelligent life. These terms include (1) the number of stars in the Milky Way that survive sufficiently long for intelligent life to evolve on planets around them; (2) the average number of planets around each of these stars; (3) the fraction of these planets with conditions suitable for life; (4) the probability that life actually arises on these suitable planets; and (5) the chance that life on such a planet evolves to produce an intelligent civilization, by which astronomers typically mean a form of life capable of communicating with ourselves. When we multiply these five terms, we obtain the number of planets in the Milky Way that possess an intelligent civilization at some point in their history. To make the Drake equation yield the number that we seek—the number of intelligent civilizations that exist at any representative time, such as the present—we must multiply this product by a sixth and final term, the ratio of the average lifetime of an intelligent civilization to the total lifetime of the Milky Way galaxy (about 10 billion years).

Each of the Drake equation's six terms requires astronomical, biological, or sociological knowledge. We now have good estimates of the equation's first two terms, and seem likely to obtain a useful estimate of the third before long. On the other hand, terms four and five—the probability that life arises on a suitable planet, and the probability that this life evolves to produce an intelligent civilization—require that we discover and examine various forms of life throughout the galaxy. For now, anyone can argue almost as well as experts can about the value of these terms. What is the probability, for example, that if a planet does have conditions suitable for life, then life will actually begin on that planet? A scientific approach to this question cries out for the study of several planets suitable for life for a few billion years to see how many do produce life. Any attempt to determine the average lifetime of a civilization in the Milky Way likewise requires several billion years of observation, once we have located a sufficiently large number of civilizations to provide a representative sample.

Isn't this a hopeless task? A full solution of the Drake equation indeed lies immensely far in the future—unless we encounter other civilizations that have already solved it, perhaps using us as a data point. But the equation nevertheless provides useful insights for what it takes to estimate how many civilizations exist in our galaxy now. The six terms in the Drake equation all resem-

ble one another mathematically in their effect on the total outcome: each of them exerts a direct, multiplying effect on the
equation's answer. If, for instance, you assume that one in three
planets suitable for life actually produces life, but later explorations reveal that this ratio actually equals 1 in 30, you will have
overestimated the number of civilizations by a factor of 10,
assuming that your estimates for the other terms prove correct.

Judging by what we now know, the first three terms in the
Drake equation imply that billions of potential sites for life exist
in the Milky Way. (We restrict ourselves to the Milky Way out of
modesty, plus our awareness that civilizations in other galaxies
will have a much more difficult time in establishing contact with
us, or we with them.) If you like, you can engage in soul-search-
ing arguments with your friends, family, and colleagues about the
value of the remaining three terms, and decide on numbers that
will provide your own estimate for the total number of technologically proficient civilizations in our galaxy. If you believe, for
example, that most planets suitable for life do produce life, and
that most planets with life do evolve intelligent civilizations, you
will conclude that billions of planets in the Milky Way produce
an intelligent civilization at some point in their time line. If, on
the other hand, you conclude that only one suitable planet in a
thousand does produce life, and only one life-bearing planet in a
thousand evolves intelligent life, you will have only thousands,
not billions, of planets with an intelligent civilization. Does this
enormous range of answers—potentially even wider than the
examples given here—imply that the Drake equation presents
wild and unbridled speculation rather than science? Not at all.
This result simply testifies to the Herculean labor that scientists,
along with everyone else, face in attempting to answer an extremely
complex question on the basis of highly limited knowledge.

The difficulty that we face in estimating the values of the last
three terms in the Drake equation highlights the treacherous step

that we take whenever we make a sweeping generalization from a single example—or from none at all. We are hard pressed, for example, to estimate the average lifetime of a civilization in the Milky Way when we do not even know how long our own will last. Must we abandon all faith in our estimates of these numbers? This would emphasize our ignorance while depriving us of the joy of speculation. If, in the absence of data or dogma, we seek to speculate conservatively, the safest course (though one that might eventually prove to be erroneous) rests on the notion that we are not special. Astrophysicists call this assumption the "Copernican principle" after Nicolaus Copernicus, who, in the mid-1500s, placed the Sun in the middle of our solar system, where it turned out to belong. Until then, despite a third-century B.C. proposal for a Sun-centered universe by the Greek philosopher Aristarchus, the Earth-centered cosmos had dominated popular opinion during most of the past two millennia. Codified by the teachings of Aristotle and Ptolemy, and by the preachings of the Roman Catholic Church, this dogma led most Europeans to accept Earth as the center of all creation. This must have appeared both self-evident from a look at the heavens and the natural result of God's plan for the planet. Even today, enormous segments of Earth's human population—quite likely a significant majority—continue to draw this conclusion from the fact that Earth seemingly remains immobile while the sky turns around us.

Although we have no guarantee that the Copernican principle can guide us correctly in all scientific investigations, it provides a useful counterweight to our natural tendency to think of ourselves as special. Even more significant is that the principle has an excellent track record so far, leaving us humbled at every turn: Earth does not occupy the center of the solar system, nor does the solar system occupy the center of the Milky Way galaxy, nor the Milky Way galaxy the center of the universe. And in case you believe that the edge is a special place, we are not at the edge of anything, either. A wise contemporary attitude therefore assumes that life on Earth likewise follows the Copernican principle. If so, how can life on Earth, its origins, and its components and structure provide clues about life elsewhere in the universe?

In attempting to answer this question, we must digest an enormous array of biological information. For every cosmic data point, gleaned by long observations of objects at enormous distances from us, we know thousands of biological facts. The diversity of life leaves us all, but especially biologists, awestruck on a daily basis. On this single planet Earth, there co-exist (among countless other life forms), algae, beetles, sponges, jellyfish, snakes, condors, and giant sequoias. Imagine these seven living organisms lined up next to each other in order of size. If you didn't know better, you would be challenged to believe that they all came from the same universe, much less the same planet. Try describing a snake to somebody who has never seen one: "You gotta believe me. I just saw this animal on planet Earth that (1) stalks its prey with infrared detectors, (2) swallows whole live animals up to five times bigger than its head, (3) has no arms or legs or any other appendage, yet (4) can slide along level ground almost as fast as you can walk!"

In contrast to the amazing variety of life on Earth, the constricted vision and creativity of Hollywood writers who imagine other forms of life is shameful. Of course, the writers probably blame a public that favors familiar spooks and invaders over truly alien ones. But with a few notable exceptions, such as the life forms in *The Blob* (1958) and in Stanley Kubrick's *2001: A Space Odyssey* (1968), Hollywood aliens all look remarkably humanoid. No matter how ugly (or cute) they may be, nearly all of them have two eyes, a nose, a mouth, two ears, a head, a neck, shoulders, arms, hands, fingers, a torso, two legs, two feet—and they can walk. From an anatomical view, these creatures are practically indistinguishable from humans, yet they are supposed to live on other planets, the products of independent lines of evolution. A clearer violation of the Copernican principle can hardly be found.

Astrobiology—the study of the possibilities for extraterrestrial life—ranks among the most speculative of sciences, but astrobiologists can already assert with confidence that life elsewhere in the universe, intelligent or otherwise, will surely look at least as exotic as some of Earth's own life forms, and quite probably more so. When we assess the chances of life elsewhere in the universe, we must attempt to shake from our brains the notions that Hollywood has implanted. Not an easy task, but essential if we hope to reach a scientific rather than an emotional estimate of our chances of finding creatures with whom we may someday have a quiet conversation.

CHAPTER 15

The Origin of Life on Earth

The search for life in the universe begins with a deep question: What is life? Astrobiologists will tell you honestly that this question has no simple or generally accepted answer. Not much use to say that we'll know it when we see it. No matter what characteristic we specify to separate living from nonliving matter on Earth, we can always find an example that blurs or erases this distinction. Some or all living creatures grow, move, or decay, but so too do objects that we would never call alive. Does life reproduce itself? So does fire. Does life evolve to produce new forms? So do certain crystals that grow in watery solutions. We can certainly say that you can tell some forms of life when you see them—who could fail to see life in a salmon or an eagle?—but anyone familiar with life in its diverse forms on Earth will admit that many creatures will remain entirely undetected until the luck of time and the skill of an expert reveal their living nature.

Since life is short, we must press onward with a rough-and-ready, generally appropriate criterion for life. Here it is: Life consists of sets of objects that can both reproduce and evolve. We shall not call a group of objects alive simply because they make more of themselves. To qualify as life, they must also evolve into new forms as time passes. This definition therefore eliminates the possibility that any single object can be judged to be alive. Instead, we must examine a range of objects in space and follow them through time. This definition of life may yet prove too restrictive, but for now we shall employ it.

As biologists have examined the different types of life on our planet, they have discovered a general property of Earthlife. The matter within every living Earth creature mainly consists of just four chemical elements: hydrogen, oxygen, carbon, and nitrogen. All the other elements together contribute less than one percent of the mass of any living organism. The elements beyond the big four include small amounts of phosphorus, which ranks as the most important, and is essential to most forms of life, together with still smaller amounts of sulfur, sodium, magnesium, chlorine, potassium, calcium, and iron.

But can we conclude that this elemental property of life on Earth must likewise describe other forms of life in the cosmos? Here we can apply the Copernican principle in full vigor. The four elements that form the bulk of life on Earth all appear on the short list of the universe's six most abundant elements. Since the other two elements on that list, helium and neon, almost never combine with anything else, life on Earth consists of the most abundant and chemically active ingredients in the cosmos. Of all the predictions that we can make about life on other worlds, the surest seems to be that their life will be made of elements nearly the same as those used by life on Earth. If life on our planet consisted primarily of four extremely rare elements in the cosmos, such as niobium, bismuth, gallium, and plutonium, we would have an excellent reason to suspect that we represent something special in the universe. Instead, the chemical composition of life on our planet inclines us toward an optimistic view of life's possibilities beyond Earth.

The composition of life on Earth fits the Copernican principle even more than one might initially suspect. If we lived on a planet made primarily of hydrogen, oxygen, carbon, and nitrogen, then the fact that life consists primarily of these four elements would hardly surprise us. But Earth is mainly made of oxygen, iron, silicon, and magnesium, and its outermost layers are mostly oxygen, silicon, aluminum, and iron. Only one of these elements, oxygen, appears on the list of life's most abundant elements. When we look into Earth's oceans, which are almost entirely hydrogen and oxygen, it is surprising that life lists carbon and nitrogen among its most abundant elements, rather than chlorine, sodium, sulfur, calcium, or potassium, which are the most common elements dissolved in seawater. The distribution of the elements in life on Earth resembles the composition of the stars far more than that of Earth itself. As a result, life's elements are more cosmically abundant than Earth's—a good start for those who hope to find life in a host of situations.

Once we have established that the raw materials for life are abundant throughout the cosmos, we may proceed to ask: How often do these raw materials, along with a site on which these materials can collect and a convenient source of energy such as a nearby star, lead to the existence of life itself? Someday, when we have made a good survey of possible sites for life in our Sun's neighborhood, we shall have a statistically accurate answer to this question. In the absence of these data, we must take a roundabout path to an answer and ask, How did life begin on Earth?

The origin of life on Earth remains locked in murky uncertainty. Our ignorance about life's beginnings stems in large part from the fact that whatever events made inanimate matter come alive occurred billions of years ago and left no definitive traces behind. For times more than 4 billion years in the past, the fossil and geological record of Earth's history does not exist. Yet the interval in solar system history between 4.6 and 4 billion years ago—the first 600 million years after the Sun and its planets had formed— includes the era when most paleobiologists, specialists in reconstructing life that existed during long-vanished epochs, believe that life first appeared on our planet.

The absence of all geological evidence from epochs more than 4 billion years ago arises from motions of Earth's crust, familiarly called continental drift but scientifically known as plate tectonics. These motions, driven by heat that wells up from Earth's interior, continually force pieces of our planet's crust to slide, collide, and ride by or over one another. Plate tectonic motions have slowly buried everything that once lay on Earth's surface. As a result, we possess few rocks older than 2 billion years, and none more than 3.8 billion years in age. This fact, together with the reasonable conclusion that the most primitive forms of life had little chance of leaving behind fossil evidence, has left our planet devoid of any reliable record of life during Earth's first 1 or 2 billion years. The oldest definite evidence we have for life on Earth takes us back "only" 2.7 billion years into the past, with indirect indications that life did exist more than 1 billion years before then.

Most paleobiologists believe that life must have appeared on Earth at least 3 billion years ago, and quite possibly more than 4 billion years ago, within the first 600 million years after Earth formed. Their conclusion relies on a reasonable supposition about primitive organisms. At times a bit less than 3 billion years ago, significant amounts of oxygen began to appear in Earth's atmosphere. We know this from Earth's geological record independently of any fossil remains: oxygen promotes the slow rusting of iron-rich rocks, which produces lovely red tones like those of the rocks in Arizona's Grand Canyon. Rocks from the pre-oxygen era show neither any such colors nor other telltale signs of the element's presence.

The appearance of atmospheric oxygen was the greatest pollution ever to occur on Earth. Atmospheric oxygen does more than combine with iron; it also takes food from the (metaphorical) mouths of primitive organisms by combining with all the simple molecules that could otherwise have provided nutrients for early forms of life. As a result, oxygen's appearance in Earth's atmosphere meant that all forms of life had to adapt or die—and that if life had not appeared by that time, it could never do so thereafter, because the would-be organisms would have nothing to eat, for their potential food would have rusted away. Evolutionary adaptation to this pollution worked well in many cases, as all oxygen-breathing animals can testify. Hiding from the oxygen also did the trick. To this day, every animal's stomach, including our own, harbor billions of organisms that thrive in the anoxic environment that we provide, but would die if exposed to air.

What made Earth's atmosphere relatively rich in oxygen? Much of it came from tiny organisms floating in the seas, which released oxygen as part of their photosynthesis. Some oxygen would have appeared even in the absence of life, as UV from sunlight broke apart some of the H_2O molecules at the ocean surfaces, releasing hydrogen and oxygen atoms into the air. Wherever a planet exposes significant amounts of liquid water to starlight, that planet's atmosphere should likewise gain oxygen, slowly but surely, over hundreds of millions or billions of years. There too, atmospheric oxygen would prevent life from originating by combining with all possible nutrients that could sustain life. Oxygen kills! Not what we usually say about this eighth element on the periodic table, but for life throughout the cosmos, this verdict appears accurate: Life must begin early in a planet's history, or else the appearance of oxygen in its atmosphere will put the kibosh on life forever.

———

By a strange coincidence, the epoch missing from the geological record that includes the origin of life also includes the so-called era of bombardment, which covers those critical first few hundred million years after Earth had formed. All portions of Earth's surface must then have endured a continual rain of objects. During those several hundred thousand millennia, infalling objects as large as the one that made the Meteor Crater in Arizona must have struck our planet several times in every century, with much larger objects, each several miles in diameter, colliding with Earth every few thousand years. Each one of the large impacts would have caused a local remodeling of the surface, so a hundred thousand impacts would have produced global changes in our planet's topography.

How did these impacts affect the origin of life? Biologists tell us that they might have triggered both the appearance and the extinction of life on Earth, not once but many times. Much of the infalling material during the era of bombardment consisted of comets, which are essentially large snowballs laden with tiny rocks and dirt. Their cometary "snow" consists of both frozen water and frozen carbon dioxide, familiarly called dry ice. In addition to their snow, grit, and rocks rich in minerals and metals, the comets that bombarded Earth during its first few hundred million years contained many different types of small molecules, such as methane, ammonia, methyl alcohol, hydrogen cyanide, and formaldehyde. These molecules, along with water, carbon monoxide, and carbon dioxide, provide the raw materials for life. They all consist of hydrogen, carbon, nitrogen, and oxygen, and they all represent the first steps in building complex molecules.

Cometary bombardment therefore appears to have provided Earth with some of the water for its oceans and with material from which life could begin. Life itself might have arrived in these comets, though their low temperatures, typically hundreds of degrees below zero Fahrenheit, argue against the formation of truly complex molecules. But whether or not life arrived with the comets, the largest objects to strike during the era of bombardment might well have destroyed life that had arisen on Earth. Life might have begun, at least in its most primitive forms, in fits and starts many times over, with each new set of organisms surviving for hundreds of thousands or even millions of years, until a collision with a particularly large object wreaked such havoc on Earth that all life perished, only to appear again, and to be destroyed again, after the passage of a similar amount of time.

We can gain some confidence in the fits-and-starts origin of life from two well-established facts. First, life appeared on our planet sooner rather than later, during the first third of Earth's lifetime. If life could and did arise within a billion years, perhaps it could do so in far less time. The origin of life might require no more than a few million, or a few tens of millions, of years. Second, we know that collisions between large objects and Earth have, at intervals of time measured in tens of millions of years, destroyed most of the species alive on our planet. The most famous of these, the Cretaceous-Tertiary extinction 65 million years ago, killed all the non-ovian dinosaurs, along with enormous numbers of other species. Even this mass extinction fell short of the most extensive one, the Permian-Triassic mass extinction, that destroyed nearly 90 percent of all species of marine life and 70 percent of all terrestrial vertebrate species, 252 million years ago, leaving fungi as the dominant forms of life on land.

The Cretaceous-Tertiary and Permian-Triassic mass extinctions arose from the collisions of Earth with objects one or two dozen miles across. Geologists have found an enormous 65-million-year-old impact crater, coincident in time with the Cretaceous-Tertiary extinction, that stretches across the northern Yucatán Peninsula and the adjoining seabed. A large crater exists with the same age as the Permian-Triassic extinction, discovered

off the northwest coast of Australia, but this mass dying might
have arisen from something in addition to a collision, perhaps
from sustained volcanic eruptions. Even the single example of
the Cretaceous-Tertiary dinosaur extinction reminds us of the
immense damage to life that the impact of a comet or asteroid can
produce. During the era of bombardment, Earth must have reeled
not only from this sort of impact, but also from the much more
serious effects of collisions with objects 50, 100, or even 250 miles
in diameter. Each of these collisions must have cleared the decks
of life, either completely or so thoroughly that only a tiny per-
centage of living organisms managed to survive, and they must
have occurred far more often than collisions with ten-mile-wide
objects do now. Our present knowledge of astronomy, biology,
chemistry, and geology points toward an early Earth ready to produce life, and a cosmic environment ready to eliminate it. And
wherever a star and its planets have recently formed, intense
bombardment by debris left over from the formation process may
even now be eliminating all forms of life on those planets.

More than 4 billion years ago, most of the debris from the solar
system's formation either collided with a planet or moved into
orbits where collisions could not occur. As a result, our cosmic
neighborhood gradually changed from a region of continual bombardment to the overall calm that we enjoy today, broken only at
multi-million-year intervals by collisions with objects large
enough to threaten life on Earth. You can compare the ancient
and ongoing threat from impacts whenever you look at the full
moon. The giant lava plains that create the face of the "man in
the Moon" are the result of tremendous impacts some 4 billion
years ago, as the era of bombardment ended, whereas the crater
named Tycho, fifty-five miles across, arose from a smaller, but still
highly signficant, impact that occurred soon after the dinosaurs
disappeared from Earth.

We do not know whether life already existed 4 billion years

ago, having survived the early impact storm, or whether life arose on Earth only after relative tranquility began. These two alternatives include the possibility that incoming objects seeded our planet with life, either during the era of bombardment or soon afterward. If life began and died out repeatedly while chaos rained down from the skies, the processes by which life originated seem robust, so that we might reasonably expect them to have occurred again and again on other worlds similar to our own. If, on the other hand, life arose on Earth only once, either as home-grown life or as the result of cosmic seeding, its origin may have occurred here by luck.

In either case, the crucial question of how life actually began on Earth, either once or many times over, has no good answer, though speculation on the subject has acquired a long and intriguing history. Great rewards lie in store for those who can resolve this mystery. From Adam's rib to Dr. Frankenstein's monster, humans have answered the question by invoking a mysterious *élan vital* that imbues otherwise inanimate matter with life.

Scientists seek to probe more deeply, with laboratory experiments and examinations of the fossil record that attempt to establish the height of the barrier between inanimate and animate matter, and to find how nature breached this dike. Early scientific discussions about the origin of life imagined the interaction of simple molecules, concentrated in pools or tide ponds, to create more complex ones. In 1871, a dozen years after the publication of Charles Darwin's marvelous book *The Origin of Species*, in which he speculated that "probably all of the organic beings which have ever lived on this Earth have descended from some one primordial form," Darwin wrote to his friend Joseph Hooker that

It is often said that all the conditions for the first production of a living organism are now present, which could ever have been present. But if (and oh! what a big if!) we could conceive in some warm little pond, with all sorts of ammonia and phosphoric salts, light, heat, electricity, &c., present, that a proteine [*sic*] compound was chemically formed ready to undergo still more complex changes, at the present day such matter would be instantly absorbed, which would not have been the case before living creatures were found.

In other words, when Earth was ripe for life, the basic compounds necessary for metabolism might have existed in surplus, with nothing in existence to eat them (and, as we have discussed, no oxygen to combine with them and spoil their chances to serve as food).

From a scientific perspective, nothing succeeds like experiments that can be compared with reality. In 1953, seeking to test Darwin's conception of the origin of life in ponds or tide pools, Stanley Miller, who was then a U.S. graduate student working at the University of Chicago with the Nobel laureate Harold Urey, performed a famous experiment that duplicated the conditions within a highly simplified and hypothetical pool of water on the early Earth. Miller and Urey partly filled a laboratory flask with water and topped the water with a gaseous mixture of water vapor, hydrogen, ammonia, and methane. They heated the flask from below, vaporizing some of the contents and driving them along a glass tube into another flask, where an electrical discharge simulated the effect of lightning. From there the mixture returned to the original flask, completing a cycle that would be repeated over and over during a few days, rather than a few thousand years. After this entirely modest time interval, Miller and Urey found the water in the lower flask to be rich in "organic gunk," a compound of numerous complex molecules, including different types of sugar, as well as two of the simplest amino acids, alanine and guanine.

Since protein molecules consist of twenty types of amino acids arranged into different structural forms, the Miller-Urey experiment takes us, in a remarkably brief time, a significant part of the way from the simplest molecules to the amino-acid molecules that form the building blocks of living organisms. The Miller-Urey experiment also made some of the modestly complex molecules called nucleotides, which provide the key structural element for DNA, the giant molecule that carries instructions for forming new copies of an organism. Even so, a long path remains before life emerges from experimental laboratories. An enormously significant gap, so far unbridged by human experiment or invention, separates the formation of amino acids—even if our experiments produced all twenty of them, which they do not—and the creation of life. Amino-acid molecules have also been found in some of the oldest and least altered meteorites, believed to have remained unchanged for nearly the entire 4.6-billion-year history of the solar system. This supports the general conclusion that natural processes can make amino acids in many different situations. A balanced view of the experimental results finds nothing totally surprising: The simpler molecules found in living organisms form quickly in many situations, but life does not. The key question still remains: How does a collection of molecules, even one primed for life to appear, ever generate life itself?

Since the early Earth had not weeks but many million years in which to bring forth life, the Miller-Urey experimental results seemed to support the tide-pool model for life's beginnings. Today, however, most scientists who seek to explain life's origin consider the experiment to have been significantly limited by its techniques. Their shift in attitude arose not from doubting the test's results, but rather from recognizing a potential flaw in the hypotheses underlying the experiment. To understand this flaw, we must consider what modern biology has demonstrated about the oldest forms of life.

————

Evolutionary biology now relies on careful study of the similarities and differences between living creatures in their molecules of DNA and RNA, which carry the information that tells an organism how to function and how to reproduce. Careful comparison of these relatively enormous and complex molecules has allowed biologists, among whom the great pioneer has been Carl Woese, to create an evolutionary tree of life that records the "evolutionary distances" between various life forms, as determined by the degrees to which these life forms have nonidentical DNA and RNA.

The tree of life consists of three great branches, Archaea, Bacteria, and Eucarya, that replace the biological "kingdoms" formerly believed to be fundamental. The Eucarya includes every organism whose individual cells have a well-defined center or nucleus that contains the genetic material governing the cells' reproduction. This characteristic makes Eucarya more complex than the other two types, and indeed every form of life familiar to the non-expert belongs to this branch. We may reasonably conclude that Eucarya arose later than Archaea or Bacteria. And because Bacteria lie farther from the origin of the tree of life than the Archaea do—for the simple reason that their DNA and RNA has changed more—the Archaea, as their name implies, almost certainly represent the oldest forms of life. Now comes a shocker: Unlike the Bacteria and Eucarya, the Archaea consist mainly of "extremophiles," organisms that love to live, and live to love, in what we now call extreme conditions: temperatures near or above the boiling point of water, high acidity, or other situations that would kill other forms of life. (Of course, if the extremophiles had their own biologists, they would classify themselves as normal and any life that thrives at room temperature as an extremophile.) Modern research into the tree of life tends to suggest that life began with the extremophiles, and only later evolved into forms of life that benefit from what we call normal conditions.

In that case, Darwin's "warm little pond," as well as the tide pools duplicated in the Miller-Urey experiment, would evaporate into the mist of rejected hypotheses. Gone would be the relatively mild cycles of drying and wetting. Instead, those who seek to find the places where life may have begun would have to look to locales where extremely hot water, possibly laden with acids, surges from Earth.

The past few decades have allowed oceanographers to discover just such places, along with the strange forms of life they support. In 1977, two oceanographers piloting a deep sea submersible vehicle discovered the first deep sea vents, a mile and a half beneath the calm surface of the Pacific Ocean near the Galápagos Islands. At these vents, Earth's crust behaves locally like a household cooker, generating high pressure inside a heavy-duty pot with a lockable lid and heating water beyond its ordinary boiling temperature without letting it reach an actual boil. As the lid partially lifts, the pressurized, superheated water spews out from below Earth's crust into the cold ocean basins.

The superheated seawater that emerges from these vents carries dissolved minerals that quickly collect and solidify to surround the vents with giant, porous rock chimneys, hottest in their cores and coolest at the edges that make direct contact with seawater. Across this temperature gradient live countless life forms that have never seen the Sun and care nothing for solar heating, though they do require the oxygen dissolved in seawater, which in turn comes from the existence of solar-driven life near the surface. These hardy bugs live on geothermal energy, which combines heat left over from Earth's formation with heat continuously produced by the radioactive decay of unstable isotopes such as aluminum-26, which lasts for millions of years, and potassium-40, which lasts for billions.

Near these vents, far below the depths to which any sunlight can penetrate, the oceanographers found tube worms as long as a man, thriving amidst large colonies of bacteria and other small creatures. Instead of drawing their energy from sunlight, as plants do with photosynthesis, life near deep sea vents relies on "chemosynthesis," the production of energy by chemical reactions, which in turn depend on geothermal heating.

How does this chemosynthesis occur? The hot water gushing from the deep sea vents emerges laden with hydrogen-sulfur and hydrogen-iron compounds. Bacteria near the vents combine these molecules with the hydrogen and oxygen atoms in water molecules, and with the carbon and oxygen atoms of the carbon dioxide molecules dissolved in sea water. These reactions form larger molecules—carbohydrates—from carbon, oxygen, and hydrogen atoms. Thus the bacteria near deep sea vents mimic the activities of their cousins far above, which likewise make carbohydrates from carbon, oxygen, and hydrogen. One set of microorganisms draws the energy to make carbohydrates from sunlight, and the other from chemical reactions at the ocean floors. Close by the deep sea vents, other organisms consume the carbohydrate-making bacteria, profiting from their energy in the same way that animals eat plants, or eat plant-eating animals.

In the chemical reactions near deep sea vents, however, more goes on than the production of carbohydrate molecules. The iron and sulfur atoms, which are not included in the carbohydrate molecule, combine to make compounds of their own, most notably crystals of iron pyrite, familiarly called "fool's gold," known to the ancient Greeks as "fire stone" because a good blow from another rock will strike sparks from it. Iron pyrite, the most abundant of all the sulfur-bearing minerals found on Earth, might have played a crucial role in the origin of life by encouraging the formation of carbohydratelike molecules. This hypothesis sprang from the mind of a German patent attorney and amateur biologist, Günter Wächtershäuser, whose profession hardly excludes him from biological speculation, any more than Einstein's work as a patent attorney barred him from insights into physics. (To be sure, Einstein had an advanced degree in physics, while Wächtershäuser's biology and chemistry are mainly self-taught.)

In 1994, Wächtershäuser proposed that the surfaces of iron pyrite crystals, formed naturally by combining iron and sulfur that surged from deep sea vents early in Earth's history, would have offered natural sites where carbon-rich molecules could accumulate, acquiring new carbon atoms from the material ejected by the nearby vents. Like those who hypothesize that life began in ponds or tide pools, Wächtershäuser has no clear way to pass from the building blocks to living creatures. Nevertheless, with his emphasis on the high-temperature origin of life, he may prove to be on the right track, as he firmly believes. Referring to the highly ordered structure of iron pyrite crystals, on whose surfaces the first complex molecules for life might have formed, Wächtershäuser has confronted his critics at scientific conferences with the striking statement that "Some say that the origin of life brings order out of chaos—but I say, 'order out of order out of order!'" Delivered with German brio, this claim acquires a certain resonance, though only time can tell how accurate it may be.

So which basic model for life's origin is more likely to prove correct—tide pools at the ocean's edge, or superheated vents on the ocean floors? For now, the betting is about even. Experts on the origin of life have challenged the assertion that life's oldest forms lived at high temperatures, because current methods for placing organisms at different points along the branches of the tree of life remain the subject of debate. In addition, computer programs that trace out how many compounds of different types existed in ancient RNA molecules, the close cousins of DNA that apparently preceded DNA in life's history, suggest that the compounds favored by high temperatures appeared only after life had undergone some relatively low-temperature history.

Thus the outcome of our finest research, as so often occurs in science, proves unsettling to those who seek certainty. Although we can state approximately when life began on Earth, we don't know where or how this marvelous event occurred. Paleobiologists have recently given the elusive ancestor of all Earthlife the name LUCA, for the last universal common ancestor. (See how firmly these scientists' minds have remained fixed to our planet: they should call life's progenitor LECA, for the last Earthly common ancestor.) For now, naming this ancestor—a set of primitive organisms that all shared the same genes—mainly underscores the distance that we still must travel before we can pierce the veil that separates life's origin from our understanding.

More than a natural curiosity as to our own beginnings hinges on the resolution of this issue. Different origins for life imply different possibilities for its origin, evolution, and survival both here and elsewhere in the cosmos. For example, Earth's ocean floors may provide the most stable ecosystem on our planet. If a jumbo asteroid slammed into Earth and rendered all surface life extinct, the oceanic extremophiles would almost certainly continue undaunted in their happy ways. They might even evolve to repopulate Earth's surface after each extinction episode. And if the Sun were mysteriously plucked from the center of the solar system and Earth drifted through space, this event would hardly merit attention in the extremophile press, as life near deep sea vents might continue relatively undisturbed. But in 5 billion years, the Sun will become a red giant as it expands to fill the inner solar system. Meanwhile, Earth's oceans will boil away and Earth itself will partially vaporize. Now that would be news for any form of Earthlife.

The ubiquity of extremophiles on Earth leads us to a profound question: Could life exist deep within many of the rogue planets or planetesimals that were ejected from the solar system during its formation? Their "geo"thermal reservoirs could last for billions of years. What about the countless planets that were forcibly ejected by every other solar system that ever formed? Could interstellar space be teeming with life—formed and evolved deep within these starless planets? Before astrophysicists recognized the importance of extremophiles, they envisioned a "habitable zone" surrounding each star, within which water or another substance could maintain itself as a liquid, allowing molecules to float, interact, and produce more complex molecules. Today, we must modify this concept, so that far from being a tidy region around a star that receives just the right amount of sunlight, a habitable zone can be anywhere and everywhere, maintained not by starlight heating but by localized heat sources, often generated by radioactive rocks. So the Three Bears' cottage was, perhaps, not a special place among fairy tales. Anybody's residence, even one of the Three Little Pigs', might contain a bowl of food at a temperature that is just right.

What a hopeful, even prescient fairy tale this may prove to be. Life, far from being rare and precious, may be almost as common as planets themselves. All that remains is for us to go find it.

CHAPTER 16

Searching for Life
in the Solar System

The possibility of life beyond Earth has created new job titles, applicable to only a few individuals but potentially capable of sudden growth. "Astrobiologists" or "bioastronomers" grapple with the issues presented by life beyond Earth, whatever forms that life may take. For now, astrobiologists can only speculate about extraterrestrial life or simulate extraterrestrial conditions, to which they either expose terrestrial life forms, testing how they may survive harsh and unfamiliar situations, or subject mixtures of inanimate molecules, creating a variant on the classic Miller-Urey experiment or a gloss on Wächtershäuser's research. This combination of speculation and experiment has led them to several generally accepted conclusions, which—to the extent that they describe the real universe—have highly significant implications. Astrobiologists now believe that the existence of life throughout the universe requires:

1. a source of energy;

2. a type of atom that allows complex structures to exist;

3. a liquid solvent in which molecules can float and interact; and

4. sufficient time for life to arise and to evolve.

On this short list, requirements (1) and (4) present only low barriers to the origin of life. Every star in the cosmos provides a source of energy, and all but the most massive 1 percent of these stars last for hundreds of millions or billions of years. Our Sun, for example, has furnished Earth with a steady supply of heat and light during the past 5 billion years, and will continue to do so for another 5 billion. Furthermore, we now see that life can exist entirely without sunlight, relying on geothermal heating and chemical reactions for its energy. Geothermal energy arises in part from the radioactivity of isotopes of elements such as potassium, thorium, and uranium, whose decay occurs over time scales measured in billions of years—a time scale comparable to the lifetime of all Sun-like stars.

On Earth, life satisfies point (2), the requirement of a structure-building atom, with the element carbon. Carbon atoms can each bind to one, two, three, or four other atoms, which makes them the crucial element in the structure of all the life we know. In contrast, hydrogen atoms can each bind to only one other atom, and oxygen to only one or two. Because carbon atoms can bind with as many as four other atoms, they form the "backbone" for all but the simplest molecules within living organisms, such as proteins and sugars.

Carbon's ability to create complex molecules has made it one of the four most abundant elements, together with hydrogen, oxygen, and nitrogen, in all forms of life on Earth. We have seen that although the four most abundant elements in Earth's crust have only one match with these four, the universe's six most abundant elements include all four of those in Earthlife, along with the inert gases helium and neon. This fact could support the hypothesis that life on Earth began in the stars, or in objects whose composition resembles those of the stars. In any case, the fact that carbon forms a relatively small fraction of Earth's surface but a large part of any living creature testifies to carbon's pivotal role in giving structure to life.

Is carbon essential to life throughout the cosmos? What about the element silicon, which often appears in science fiction novels as the basic structural atom for exotic forms of life? Like carbon, silicon atoms bond with as many as four other atoms, but the nature of these bonds leaves silicon far less likely than carbon to provide the structural basis for complex molecules. Carbon bonds to other atoms rather weakly, so that carbon-oxygen, carbon-hydrogen, and carbon-carbon bonds, for example, break with relative ease. This allows carbon-based molecules to form new types as they collide and interact, an essential part of any life form's metabolic activity. In contrast, silicon bonds strongly to many other types of atoms, and in particular to oxygen. Earth's crust consists largely of silicate rocks made primarily of silicon and oxygen atoms, bound together with sufficient strength to last for millions of years, and therefore unavailable to participate in forming new types of molecules.

The difference between the way that silicon and carbon atoms bond to other atoms argues strongly that we may expect to find most, if not all, extraterrestrial life forms built, as we are, with carbon, not silicon, backbones for their molecules. Other than carbon and silicon, only relatively exotic types of atoms, with cosmic abundances much lower than those of carbon or silicon, can bond to as many as four other atoms. Purely on numerical grounds, the possibility that life uses atoms such as germanium in the same way that Earthlife uses carbon seems highly remote.

Requirement number (3) specifies that all forms of life need a liquid solvent in which molecules can float and interact. The word "solvent" emphasizes that a liquid allows this float-and-interact situation, in what chemists call a "solution." Liquids allow relatively high concentrations of molecules but do not place tight restrictions on their motions. In contrast, solids lock atoms and molecules in place. They actually can collide and interact, but they do so far more slowly than in liquids. In gases, molecules will move even more freely than in liquids, and can collide with even less hindrance, but their collisions and interactions occur far less often than they do in liquids, because the density within a liquid typically exceeds that within a gas by a factor of 1,000 or more. "Had we but world enough and time," as Andrew Marvell wrote, we might find life originating in gases rather than liquids. In the real cosmos, only 14 billion years old, astrobiologists do not expect to find life that began in gas. Instead, they expect all extraterrestrial life, like all life on Earth, to consist of sacs of liquid, within which complex chemical processes occur as different types of molecules collide and form new types.

Must that liquid be water? We live on a watery planet whose oceans cover nearly three quarters of the surface. This makes us unique in our solar system, and possibly a highly unusual planet anywhere in our Milky Way galaxy. Water, which consists of molecules made from two of the most abundant elements in the cosmos, appears at least in modest amounts in comets, in meteoroids, and in most of the Sun's planets and their moons. On the other hand, liquid water in the solar system exists only on Earth and beneath the icy surface of Jupiter's large moon Europa, whose worldwide covered ocean remains only a likelihood, not a verified reality. Could other compounds offer better chances for liquid seas or ponds, within which molecules could have found their way to life? The three most abundant compounds that can remain liquid within a significant range of temperatures are ammonia, ethane, and methyl alcohol. Ammonia molecules each consist of three hydrogen atoms and one nitrogen atom, ethane of two hydrogen atoms and two carbon atoms, and methyl alcohol of four hydrogen atoms, one carbon atom, and one oxygen atom. When we consider the possibilities for extraterrestrial life, we may reasonably consider creatures that use ammonia, ethane, or methyl alcohol in the way that Earth life employs water—as the fundamental liquid within which life presumably originated, and which supplies the medium within which molecules can float their way to glory. The Sun's four giant planets possess enormous amounts of ammonia, along with smaller amounts of methyl alcohol and ethane, and Saturn's large moon Titan may well have lakes of liquid ethane on its frigid surface.

The choice of a particular type of molecule as life's basic liquid immediately implies another requirement for life: the substance must remain liquid. We would not expect life to originate in the Antarctic ice cap, or in clouds rich in water vapor, because we need liquids to allow abundant molecular interactions. Under atmospheric pressures like those at Earth's surface, water remains liquid between 0 and 100 degrees Celsius (32 to 212 degrees Fahrenheit). All three of the alternative types of solvents remain liquid within temperature ranges that extend far below water's. Ammonia, for example, freezes at −78 degrees Celsius and vaporizes at −33 degrees. This prevents ammonia from providing a liquid solvent for life on Earth, but on a world with a temperature 75 degrees colder than ours, where water could never serve as a solvent for life, ammonia might well be the charm.

21: This expanding region of gas, named IC 443 by astronomers, is the remnant of a supernova, about 5,000 light-years from the solar system. The star exploded about 30,000 years before the supernova remnant produced the light recorded in this image, obtained with the Canada-France-Hawaii Telescope at the Mauna Kea Observatory.

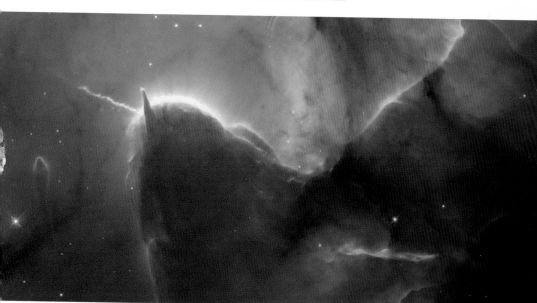

22: These wisps of gas in the Trifid nebula, about 5,000 light-years away, were imaged by the high-resolution optics of the Hubble Space Telescope. The gas in these pillars must be denser than their surroundings, which have been stripped away by radiation from young, hot stars nearby.

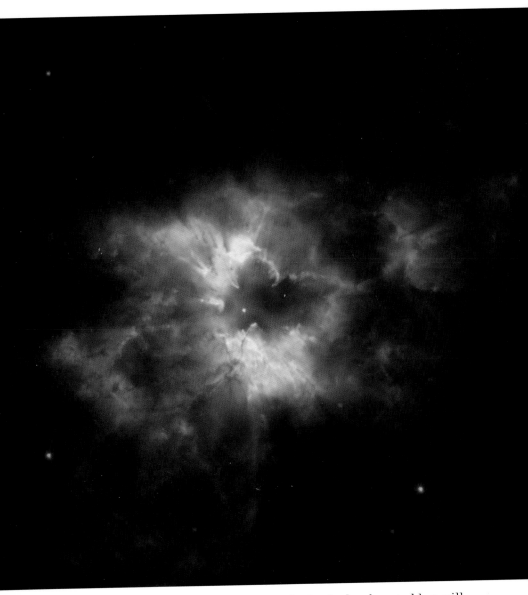

23: This nebula, called NGC 2440, surrounds the fuel-exhausted but still hot core of what was once a star. This "white dwarf" appears as a bright spot of light near the center of the nebula in this Hubble Space Telescope image. Before long, the gas surrounding this object, about 3,500 light-years from the solar system, will evaporate into space, leaving the white dwarf isolated as it slowly cools and grows dimmer.

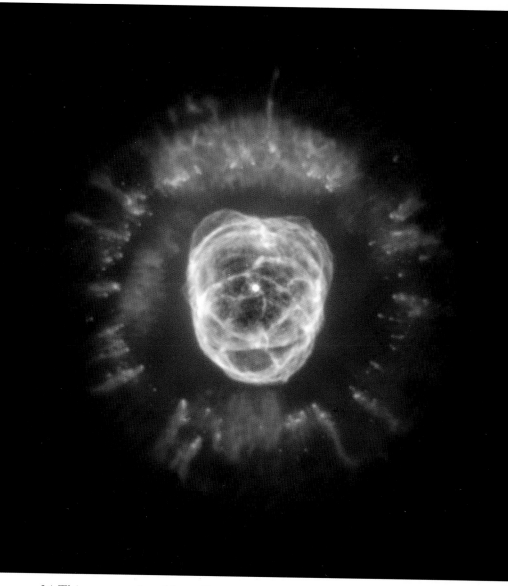

24: This spectacular object, discovered by the famous astronomer William Herschel in 1787, bears the name Eskimo nebula for its resemblance to a face surrounded by the furry hood of a parka. The nebula, about 3,000 light-years away, consists of gas expelled from an aging star and illuminated by ultraviolet radiation from that star, whose surface has grown so hot that it emits more ultraviolet than visible light. Like Herschel, astronomers call objects like these "planetary nebulae" because a small telescope shows them only as featureless disks, similar to the images of planets. This Hubble Space Telescope image removes the confusion by revealing a host of detail in the gases expanding away from the central star.

25: Amidst a star-forming region in our galaxy, a relatively cool and dense cloud of gas and dust absorbs starlight, creating the aptly named Horsehead nebula, photographed with the Canada-France-Hawaii Telescope at the Mauna Kea Observatory. This dust cloud, about 1,500 light-years from the solar system, forms part of a much larger dark and cool interstellar cloud, some of which creates the dark area below the horse's head.

26: This wide-angle photograph, taken by amateur astronomer Rick Scott in 2003, shows the bright streak produced by one of the meteors observed during the annual Perseid meteor shower in mid-August, a time when Earth encounters more space debris than usual. Moving at many miles per second, each piece of debris plows through Earth's atmosphere to the point that the meteoroid vaporizes, either partially or totally. In this photograph, the Andromeda galaxy (left of the middle) can be seen at a distance about 1 million trillion times greater than the altitude of the meteor, approximately 40 miles above Earth's surface.

27: Saturn, the Sun's second largest planet, has a beautiful system of rings, photographed in all their glory by the Hubble Space Telescope. Like the more modest ring systems around Jupiter, Uranus, and Neptune, Saturn's rings consist of swarms of millions of small particles orbiting the planet.

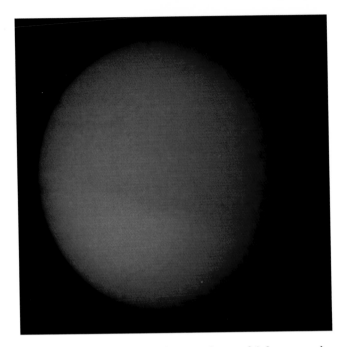

28a & b: Titan, the largest moon of Saturn, has a thick atmosphere made mainly of nitrogen molecules, but also rich in smoglike particles that permanently block its surface from view in visible light (upper image, photographed by the *Voyager 2* spacecraft in 1981). Observed in its infrared radiation, however (lower image, taken with the Canada-France-Hawaii Telescope at the Mauna Kea Observatory), Titan reveals the outlines of surface features that may well be liquid pools, areas of rock, and even glaciers of frozen hydrocarbons.

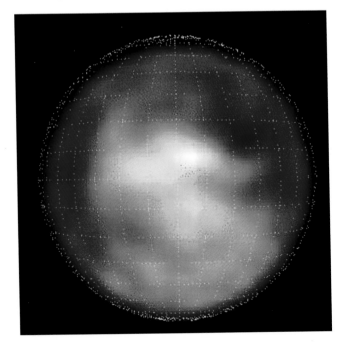

29: In December 2000, as the Cassini spacecraft passed by Jupiter en route to its Saturn rendezvous in 2004, it photographed the outer layers of the Sun's largest planet. Jupiter consists of a solid core, surrounded by gaseous layers tens of thousands of miles thick. These gases, which are mainly compounds of hydrogen with carbon, nitrogen, and oxygen, swirl in colorful patterns as the result of Jupiter's rapid rotation. The smallest features visible in this photograph are about forty miles across.

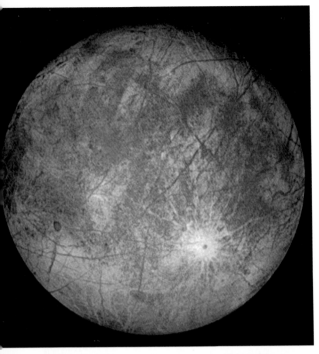

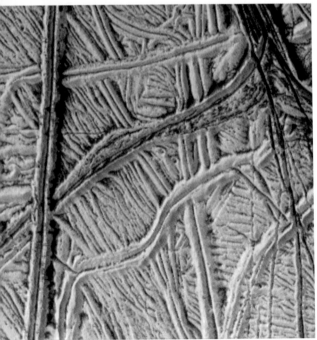

30a & b: Europa, one of Jupiter's four large moons, has about the same diameter as our Moon, but its surface displays long, straight lines that may represent worldwide cracks in its icy surface (top panel). Having secured this global view of Europa, the *Galileo* spacecraft went in for a closer inspection (bottom panel) from a distance of only 350 miles. This close-up of Europa's surface shows ice hills and straight rills, with what may be darker impact craters among them. Speculation runs high that Europa's surface ice layer, perhaps as much as half a mile thick, may cover a moonwide ocean, capable of supporting primitive forms of life.

31: During the early 1990s, radio waves from the *Magellan* spacecraft orbiting Venus, which can penetrate the planet's optically opaque atmosphere, allowed astronomers to produce this radar image of Venus' surface. Numerous large craters appear in this image, while the broad bright-colored area is the largest of Venus' highlands.

32: In 1971, the *Apollo 15* astronauts used the first vehicle on another world to explore the lunar highlands, searching for clues to the Moon's origin.

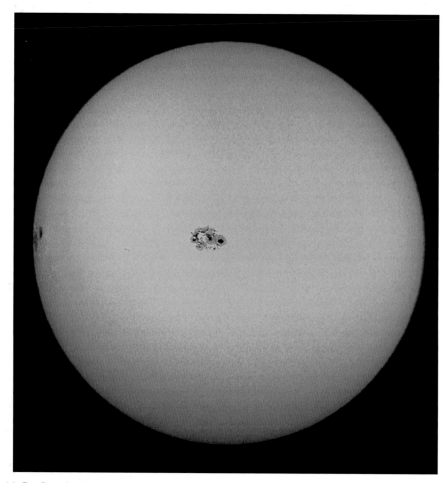

33: In October 2003, two large groups of sunspots, each several times larger than Earth, appeared on the face of the Sun, captured here by amateur astronomer Juan Carlos Casado. Rotating along with our star, these sunspots take nearly a month to cross the sun's surface and come back around again, typically fading away in about that time span. Sunspots owe their relative darkness to their cooler temperatures (about 8,000° F. in comparison to the Sun's average surface temperature of 10,000° F.). The lower temperatures arise from the influence of magnetic fields, which are also associated with violent solar eruptions, capable of emitting streams of charged particles that affect radio communications on Earth and the health of astronauts.

34: This image of Mars, taken by the Hubble Space Telescope during the planet's close approach to Earth in 2003, shows the south polar cap (mostly frozen carbon dioxide) at the bottom. At the lower right, the large circular feature is called the Hellas impact basin. Many smaller craters dot the lighter-colored Martian highlands, while the large darker areas are the lowlands of Mars.

Columbia Hills Complex

| Anderson Hill 95.2° Azimuth 3.1 Kilometers | Brown Hill 97.4° Azimuth 2.9 Kilometers | Chawla Hill 100.8° Azimuth 3.0 Kilometers | Clark Hill 106.1° Azimuth 3.0 Kilometers | Husband Hill 113.9° Azimuth 3.1 Kilometers | Mc Cool Hill 125.1° Azimuth 4.2 Kilometers | Ramon Hill 129.7° Azimuth 4.4 Kilometers |

35: This photograph of the Martian surface, taken by the *Spirit* rover in January 2004, shows hills on the horizon a few miles away. NASA has now named seven of these hills in honor of the astronauts who died in the *Columbia* shuttle disaster on February 1, 2003. Like the two sites where the *Viking* spacecraft landed in 1976, the locations where the *Spirit* and *Opportunity* rovers touched down in 2004 show rock-strewn plains with no visible signs of life.

36: A close-up view of the immediate surroundings of the *Spirit* rover shows what may be ancient bedrock, as well as younger rocks rich in compounds that on Earth typically form underwater. The prevailing reddish hue comes from iron oxides (rust) in the surface rocks and soils.

37: UCLA Professor of Biology Ken Nealson on location with one of the authors (NDT) in Death Valley during the shooting of the PBS NOVA special *Origins*. As an expert in geologically stressed microorganisms, Nealson knows that this hot, arid, and otherwise hostile environment serves as a thriving ecosystem for bacteria that live just fine within the cracks of rocks, or on their underside, shadowed from the oppressive sunlight. The reddish hue of Death Valley rocks greatly resembles that of the Martian surface.

38: Bad Day on Earth. A view by space artist Don Davis of the collision
between an asteroid and Earth 65 million years ago, which precipitated the
extinction of non-avian dinosaurs as well as 70 percent of land species,
including all animals larger than a breadbox. The ecological niches left
vacant by the dinosaurs' demise enabled mammals to evolve from tree
shrews—that had been nothing more than dino-hors d'oevres—to the
many and varied mammal forms we see today.

39: This "black smoker" rock formation, shown in vertical cross section, was hauled from the Pacific Ocean's Juan de Fuca Ridge, and now sits on display in the Hall of Planet Earth at the American Museum of Natural History in New York. Along mid-ocean ridges, water can seep through the crust and become superheated, dissolving minerals along the way. Wherever the water spews back into the ocean bottom, we find chimneylike structures, formed by the precipitation of minerals from the cooling water. The porosity of these structures, and the chemical and temperature gradients they sustain, allow an entire ecosystem to thrive on geothermal and geochemical energy sources, without regard to the Sun as a source of life-sustaining energy. The newly discovered hardiness of some forms of bacteria and other life forms on Earth has expanded the list of environments where we may hope to find life in the universe.

40: Dr. Seth Shostak, of the SETI Insitute (Search for Extra-Terrestrial Intelligence), and one of the authors (NDT) take a moment to pose between takes of *Origins* on location at the Arecibo Radio Telescope in Puerto Rico. Shostak used this largest telescope in the world to "listen" for possible intelligent signals produced by distant civilizations. The Arecibo telescope sits in a natural limestone crater. Shostak and Tyson were filmed walking and talking under the wire-mesh dish—itself an otherworldly environment.

Water's most significant distinguishing feature does not consist of its well-earned badge of "universal solvent," about which we learned in chemistry class, nor of the wide temperature range over which water remains liquid. Water's most remarkable attribute resides in the fact that while most things—water included—shrink and become denser as they cool, water that cools below 4 degrees Celsius expands, becoming progressively less dense as the temperature falls toward zero. And then, when water freezes at 0 degrees Celsius, it turns into an even less dense substance than liquid water. Ice floats, which is very good news for fish. During the winter, as the temperature of the outside air drops below freezing, 4-degree water sinks to the bottom and stays there, because it is denser than the colder water above, while a floating layer of ice builds extremely slowly on the surface, insulating the warmer water below.

Without this density inversion below 4 degrees, ponds and lakes would freeze from the bottom up, not from the top down. Whenever the outside air temperature fell below freezing, a pond's upper surface would cool and sink to the bottom as warmer water rose from below. This forced convection would rapidly drop the water's temperature to zero degrees as the surface began to freeze. Then denser, solid ice would sink to the bottom. If the entire body of water did not freeze from the bottom upward in a single season, the accumulation of ice at the bottom would allow full freezing to occur over the course of many years. In such a world, the sport of ice fishing would yield even fewer results than it does now, because all the fish would be dead—fresh-frozen. Ice anglers would find themselves on a layer of ice that was either submerged below all remaining liquid water or atop a completely frozen body of water. No longer would you need icebreakers to traverse the frozen Arctic—either the entire Arctic Ocean would be frozen solid, or the frozen parts would all have sunk to the bottom and you could just sail your ship without incident. You could slip and slide on lakes and ponds without fear of falling through. In this altered world, ice cubes and icebergs would sink, so that in April 1912, the *Titanic* would have steamed safely into the port of New York City, unsinkable (and unsunken) as advertised.

On the other hand, our mid-latitude prejudice may be showing here. Most of Earth's oceans are in no danger of freezing, whether from the top down or the bottom up. If ice sank, the Arctic Ocean might become solid, and the same might happen to the Great Lakes and the Baltic Sea. This effect could have made Brazil and India greater world powers, at the expense of Europe and the United States, but life on Earth could have persisted and flourished just as well.

Let us, for the time being, adopt the hypothesis that water has such significant advantages over its chief rivals, ammonia and methyl alcohol, that most, if not all, forms of extraterrestrial life must rely on the same solvent that Earthlife does. Armed with this supposition, along with the general abundance of the raw materials for life, the prevalence of carbon atoms, and the long stretches of time available for life to appear and to evolve, let us take a tour of our neighbors, recasting the age-old question, Where's the life? into the more modern one, Where's the water?

If you were to judge matters by the appearance of some dry and unfriendly-looking places in our solar system, you might conclude that water, while plentiful on Earth, ranks as a rare commodity elsewhere in our galaxy. But of all the molecules that can be formed with three atoms, water is by far the most abundant, largely because water's two constituents, hydrogen and oxygen, occupy positions one and three on the abundance list. This suggests that rather than asking why some objects have water, we should ask why they don't all possess large amounts of this simple molecule.

How did Earth acquire its oceans of water? The Moon's near-pristine record of craters tells us that impacting objects have struck the Moon throughout its history. We may reasonably expect that Earth has likewise undergone many collisions. Indeed, Earth's larger size and stronger gravity imply that we should have been struck many more times, and by larger objects, than the Moon. So it has been, from its birth all the way to the present. After all, Earth didn't hatch from an interstellar void, springing into existence as a preformed spherical blob. Instead, our planet grew within the condensing gas cloud that formed the Sun and its other planets. In this process, Earth grew by accreting enormous numbers of small solid particles, and eventually through incessant impacts from mineral-rich asteroids and water-rich comets. How incessant? The early impact rate of comets may have been sufficiently large to have brought us the water in all our oceans. Uncertainties (and controversies) continue to surround this hypothesis. The water that we observed in comet Halley has far greater amounts than Earth does of deuterium, an isotope of hydrogen that packs an extra neutron into its nucleus. If Earth's oceans arrived in comets, then those that hit Earth soon after the solar system formed must have had a chemical composition notably different from today's comets, or at least different from the class of comet from which Halley is drawn.

In any case, when we add their contribution to the water vapor spewn into the atmosphere by volcanic eruptions, we have no shortage of pathways by which Earth could have acquired its supply of surface water.

If you seek a waterless, airless place to visit, you need look no farther than Earth's Moon. The Moon's near-zero atmospheric pressure, combined with its two-week-long days when the temperature rises to 200 degrees Fahrenheit, causes any water to evaporate swiftly. During the two-week lunar night, the temperature can drop to 250 degrees below zero, sufficient to freeze practically anything. The *Apollo* astronauts who visited the Moon therefore brought all the water and air (and the air conditioning) that they needed for their round-trip journey.

It would be odd, however, if Earth had acquired a great deal of water, while the nearby Moon got almost none. One possibility, certainly true at least in part, is that water evaporated from the Moon's surface much more readily than from Earth's because of the Moon's lesser gravity. Another possibility suggests that lunar missions may eventually not need to import water or the assortment of products derived from it. Observations by the *Clementine* lunar orbiter, which carried an instrument to detect the neutrons produced when fast-moving interstellar particles collide with hydrogen atoms, support a long-held contention that deep-frozen ice deposits may lurk beneath craters near the Moon's north and south poles. If the Moon receives an average number of impacts per year from interplanetary flotsam, then the mixture of these impactors should, from time to time, include sizable water-rich comets, like those that strike Earth. How big could these comets be? The solar system contains plenty of comets that could melt into a puddle the size of Lake Erie.

While we can't expect a freshly laid lake to survive many sun-baked lunar days at temperatures of 200 degrees, any comet that happened to crash in the bottom of a deep crater near one of the Moon's poles (or happened to make a deep polar crater itself) would remain shrouded in darkness, because deep craters near its poles are the only places on the Moon where the "Sun don't shine." (If you thought that the Moon has a perpetual dark side, you have been badly misled by many sources, probably including Pink Floyd's 1973 album *Dark Side of the Moon*.) As light-starved Arctic and Antarctic dwellers know, the Sun in those regions never rises high in the sky at any time of day or any season of the year. Now imagine living at the bottom of a crater whose rim rises higher than the highest altitude that the Sun ever reaches. With no air to scatter sunlight into the shadows, you would live in eternal darkness.

But even in cold darkness, ice slowly evaporates. Just look at the cubes in your freezer's ice tray upon your return from a long vacation: their sizes will be distinctly smaller than when you departed. However, if ice has been well mixed with solid particles (as occurs in a comet), it can survive for thousands and millions of years at the bottom of the Moon's deep polar craters. Any outpost that we might establish on the Moon would benefit greatly from being located near this lake. Apart from the obvious advantages of having ice to melt, to filter, and then to drink, we could also profit by dissociating the water's hydrogen from its oxygen atoms. We could use the hydrogen, plus some of the oxygen, as active ingredients for rocket fuel, while keeping the rest of the oxygen for breathing. And in our spare time between space missions, we might choose to go skating.

Although Venus has nearly the same size and mass as Earth, several attributes distinguish our sister planet from all the other planets in the solar system, notably including its highly reflective, thick, dense, carbon dioxide atmosphere, which exerts a hundred times the surface pressure of Earth's atmosphere. Except for bottom-dwelling marine creatures that live at similar pressures, all forms of Earthlife would be crushed to death on Venus. But Venus' most peculiar feature resides in the relatively young craters uniformly scattered over its surface. This innocuous-sounding description implies that a recent planetwide catastrophe reset the cratering clock—and thus our ability to date a planet's surface by its buildup of craters—by wiping out the evidence of all previous impacts. A major erosive weather phenomenon such

as a planetwide flood might also have done this. But so could planetwide geologic (should we say Venusologic?) activity, such as lava
flows, which could have turned Venus' entire surface into the
American automotive dream—a totally paved planet. Whatever
events reset the cratering clock must have ceased abruptly. But
important questions remain, in particular about Venus' water. If a
planetwide flood did occur on Venus, where has all the water
gone? Did it sink below the surface? Did it evaporate into the
atmosphere? Or did the flood consist of a common substance other
than water? Even if no flood occurred, Venus presumably acquired
about as much water as its sister planet Earth. What has happened
to it?

The answer seems to be that Venus lost its water by growing too
hot, a result attributable to Venus' atmosphere. Although carbon
dioxide molecules let visible light pass by, they trap infrared radiation with great efficiency. Sunlight can therefore penetrate
Venus' atmosphere, even though atmospheric reflection reduces
the amount of sunlight that reaches the surface. This sunlight
heats the planet's surface, which radiates infrared, and which cannot escape. Instead, the carbon dioxide molecules trap it, as the
infrared radiation heats the lower atmosphere and the surface
below. Scientists call this trapping of infrared radiation the
"greenhouse effect" by loose analogy to their glass windows,
which admit visible light but block some of the infrared. Like
Venus and its atmosphere, Earth produces a greenhouse effect,
essential for many forms of life, that raises our planet's temperature by about 25 degrees Fahrenheit over what we would find in
the absence of an atmosphere. Most of our greenhouse effect
arises from the combined effects of water and carbon dioxide
molecules. Since Earth's atmosphere has only one ten-thousandth
as many carbon dioxide molecules as the atmosphere of Venus
does, our greenhouse effect pales in comparison. Nevertheless, we
continue to add more carbon dioxide by burning fossil fuels, so we

steadily increase the greenhouse effect, performing an unintended global experiment to see just what deleterious effects arise from the additional trapping of heat. On Venus, the atmospheric greenhouse effect, produced entirely by carbon dioxide molecules, raises the temperature by hundreds of degrees, giving Venus' surface furnacelike temperatures close to 500° Celsius (900° Fahrenheit)—the hottest in the solar system.

How did Venus reach this sorry state? Scientists apply the apt term "runaway greenhouse effect" to describe what happened as the infrared radiation trapped by Venus' atmosphere raised the temperatures and encouraged liquid water to evaporate. The additional water in the atmosphere trapped infrared even more effectively, increasing the greenhouse effect; this in turn caused even more water to enter the atmosphere, ratcheting up the greenhouse effect still farther. Near the top of Venus' atmosphere, solar UV radiation would break the water molecules apart into hydrogen and oxygen atoms. Because of the high temperatures, the hydrogen atoms would escape, while the heavier oxygen combined with other atoms, never to form water again. With the passage of time, all the water that Venus once had on or near its surface has been essentially baked out of the atmosphere and lost to the planet forever.

Similar processes occur on Earth, but at a much lower rate because we have much lower atmospheric temperatures. Our mighty oceans now comprise most of Earth's surface area, though their modest depth gives them only about one five-thousandth of Earth's total mass. Even this small fraction of the total allows the oceans to weigh in at a hefty 1.5 quintillion tons, 2 percent of which is frozen at any given time. If Earth should ever undergo a runaway greenhouse effect like the one that has occurred on Venus, our atmosphere would trap larger amounts of solar energy, raising the air temperature and making the oceans evaporate swiftly into the atmosphere as they sustained a rolling boil. This would be bad news. Apart from the obvious ways that Earth's flora and fauna would die, an especially pressing cause of death would result from Earth's atmosphere growing three hundred times more massive as it thickens with water vapor. We would be crushed and baked by the air we breathe.

Our planetary fascination (and ignorance) are hardly limited to Venus. With its long dry, still preserved meandering riverbeds, floodplains, river deltas, networks of tributaries, and river-eroded canyons, Mars must once have been a primeval Eden of water in motion. If any place in the solar system other than Earth ever boasted a flourishing water supply, it was Mars. For reasons unknown, however, today Mars has a bone-dry surface. Close examination of Venus and Mars, our sister and brother planets, forces us to look at Earth anew and to wonder how fragile our surface supply of liquid water may turn out to be.

Early in the twentieth century, imaginative observations of Mars by the noted American astronomer Percival Lowell led him to suppose that colonies of resourceful Martians had built an elaborate network of canals in order to redistribute water from Mars' polar ice caps to the more populated middle latitudes. To explain what he thought he saw, Lowell imagined a dying civilization that was exhausting its supply of water, like Phoenix discovering that the Colorado River has its limits. In his thorough yet curiously misguided treatise entitled *Mars as the Abode of Life*, published in 1909, Lowell lamented the imminent end of the Martian civilization that he imagined he saw.

Indeed, Mars seems certain to dry up to the point that its surface can support no life at all. Slowly but surely, time will snuff life out, if it has not done so already. When the last living ember dies away, the planet will roll on through space as a dead world, its evolutionary career forever ended.

Lowell happened to get one thing right. If Mars ever had a civilization (or any kind of life at all) that required water on the surface, it must have faced catastrophe, because at some unknown time in Martian history, and for some unknown reason, all the surface water did dry up, leading to the exact fate for life—though in the past, not the present—that Lowell described. What happened to the water that flowed abundantly over Mars' surface billions of years ago remains an outstanding mystery among planetary geologists. Mars does have some water ice in its polar caps, which consist mainly of frozen carbon dioxide ("dry ice"), and a tiny amount of water vapor in its atmosphere. Although the polar caps contain the only significant amounts of water now known to exist on Mars, their total content of ice falls far below the amount needed to explain the ancient records of flowing water on Mars' surface.

If most of Mars' ancient water did not evaporate into space, its most likely hiding place lies underground, with the water trapped in the planet's subsurface permafrost. The evidence? Large craters on the Martian surface are more likely than small craters to exhibit dried mud spills over their rims. If the permafrost lies deep underground, to reach it would require a large collision. The deposit of energy from such an impact would melt this subsurface ice upon contact, causing it to splash upward. Craters with this mud-spill signature are more common in the cold, polar latitudes—just where we might expect the permafrost layer to be closer to the Martian surface. According to optimistic estimates of the Martian permafrost's ice content, the melting of Mars' subsurface layers would release enough water to give Mars a planetwide ocean tens of meters deep. A thorough search for contemporary (or fossil) life on Mars must include a plan to search in many locations, especially below the Martian surface. So far as the chance of finding life on Mars is concerned, the great question to be resolved asks, Does liquid water now exist anywhere on Mars?

Part of the answer leaps from our knowledge of physics. No liquid water can exist on the Martian surface, because the atmospheric pressure there, less than 1 percent of the value on the surface of Earth, does not allow it. As enthusiastic mountaineers know, water vaporizes at progressively lower temperatures as the atmospheric pressure decreases. At the summit of Mount Whitney, where the air pressure falls to half of its sea-level value, water boils not at 100 but at 75 degrees Celsius. On top of Mount Everest, with air pressure only a quarter of its sea-level value, boiling occurs at about 50 degrees. Twenty miles high, where the atmospheric pressure equals only 1 percent of what you feel on the sidewalks of New York, water boils at about 5 degrees Celsius. Rise a few miles higher, and liquid water will "boil" at 0 degrees—that is, it will vaporize as soon as you expose it to the air. Scientists use the word "sublimation" to describe the passage of a substance from solid to gas without any intervening liquid stage. We all know sublimation from our youth, when the ice cream man opened his magic door to reveal not only the delicacies inside but also the chunks of "dry" ice that kept them cold. Dry ice offers the ice cream man a great advantage over familiar water ice: It sublimates from solid to gas, leaving no messy liquid to clean up. An old detective story conundrum describes the man who hanged himself by standing on a cake of dry ice until it sublimated, leaving him suspended by a noose, and the detectives without a clue (unless they carefully analyzed the atmosphere in the room) as to how he did it.

What happens to carbon dioxide on Earth's surface happens to water on the surface of Mars. No chance for liquid exists there, even though the temperature on a warm day of the Martian summer rises well above 0 degrees Celsius. This seems to draw a sad veil over the prospects for life—until we realize that liquid water could exist beneath the surface. Future missions to Mars, intimately bound up with the possibility of finding ancient or even modern life on the red planet, will direct themselves toward regions where they can drill into the Martian surface in a search for the flowing elixir of life.

Elixir though it may appear, water represents a deadly substance among the chemically illiterate, to be avoided sedulously. In 1997, Nathan Zohner, a fourteen-year-old student at Eagle Rock Junior High School in Idaho, conducted a now famous (among science popularizers) science fair experiment to test antitechnology sentiments and associated chemical phobia. Zohner invited people to sign a petition that demanded either strict control or a total ban of dihydrogen monoxide. He listed some of the odious properties of this colorless and odorless substance:

- It is a major component in acid rain
- It eventually dissolves almost anything it comes in contact with
- It can kill if accidentally inhaled
- It can cause severe burns in its gaseous state
- It has been found in tumors of terminal cancer patients.

Forty-three out of fifty people approached by Zohner signed the petition, six were undecided, and one was a great supporter of the molecule and refused to sign. Yes, 86 percent of the passersby voted to ban dihydrogen monoxide (H_2O) from the environment.

Maybe that's what really happened to the water on Mars.

Venus, Earth, and Mars together provide an instructive tale about the pitfalls and payoffs from focusing on water (or possibly other solvents) as the key to life. When astronomers considered where they might find liquid water, they originally concentrated on planets that orbit at the proper distances from their host stars to maintain water in liquid form—not too close in and not too far out. Thus we begin with the tale of Goldilocks.

Once upon a time—somewhat more than 4 billion years ago—the formation of the solar system was nearly complete. Venus had formed sufficiently close to the Sun for the intense solar energy to vaporize what might have been its water supply. Mars formed so far away that its water supply became forever frozen. Only one planet, Earth, had a distance "just right" for water to remain a liquid, and whose surface would therefore become a haven for life. This region around the Sun where water can remain liquid came to be known as the habitable zone.

Goldilocks liked things "just right," too. One of the bowls of porridge in the Three Bears' cottage was too hot. Another was too cold. The third was just right, so she ate it. Upstairs, one bed was too hard. Another was too soft. The third was just right, so Goldilocks slept in it. When the Three Bears came home, they discovered not only missing porridge but also Goldilocks fast asleep in their bed. (Don't remember how the story ends, but it remains a mystery to us why the Three Bears—omnivorous and occupying the top of the food chain—did not eat Goldilocks instead.)

The relative habitability of Venus, Earth, and Mars would intrigue Goldilocks, though the actual history of these planets is somewhat more complicated than three bowls of porridge. Four billion years ago, leftover water-rich comets and mineral-rich asteroids were still pelting the planetary surfaces, although at a much lower rate than before. During this game of cosmic billiards, some planets had migrated inward from where they had formed while others were kicked into larger orbits. And among the dozens of planets that had formed, some moved on unstable orbits and crashed into the Sun or Jupiter. Others were ejected from the solar system altogether. In the end, the few planets that remained had orbits that were "just right" to survive billions of years.

Earth settled into an orbit with an average distance of 93 million miles from the Sun. At this distance, Earth intersects a measly one two-billionth of the total energy radiated by the Sun. If you assume that Earth absorbs all the energy received from the Sun, then our home planet's average temperature should be about 280 degrees Kelvin (45° F), which falls midway between winter and summer temperatures. At normal atmospheric pressures, water freezes at 273 degrees Kelvin and boils at 373 degrees, so we are well positioned with respect to the Sun for nearly all of Earth's water to remain happily in its liquid state.

Not so fast. In science you can sometimes get the right answer for the wrong reasons. Earth actually absorbs only two thirds of the energy that reaches it from the Sun. The rest is reflected back into space by Earth's surface (especially by the oceans) and by its clouds. If we factor this reflection into the equations, the average temperature for Earth drops to about 255 degrees Kelvin, well below the freezing point of water. Something must be operating to raise our average temperature to something a little more comfortable.

But wait once more. All theories of stellar evolution tell us that 4 billion years ago, when life was forming out of Earth's primordial soup, the Sun was a third less luminous than it is today, which would have left Earth's average temperature even farther below freezing. Perhaps Earth in the distant past was simply closer to the Sun. Once the early period of heavy bombardment had ended, however, no known mechanisms could have shifted stable orbits back and forth within the solar system. Perhaps the greenhouse effect from Earth's atmosphere was stronger in the past. We don't know for sure. What we do know is that habitable zones, as originally conceived, have only peripheral relevance to whether life may exist on a planet within them. This has become evident from the fact that we cannot explain Earth's history on the basis of a simple habitable-zone model, and even more from the realization that water or other solvents need not depend on the heat from a star to remain liquid.

Our solar system contains two good reminders that the "habitable-zone approach" to looking for life has severe limitations. One of them lies outside the zone where the Sun can keep water liquid, yet nevertheless has a worldwide ocean of water. The other, far too cold for liquid water, offers the possibility of another liquid solvent, poison to us but potentially prime for other forms of life. Before long we should have the opportunity to investigate both of these objects with close-up robot explorers. Let's check out what we know now about Europa and Titan.

Jupiter's moon Europa, which has about the same size as our Moon, shows crisscrossing cracks in the surface that change on time scales of weeks or months. To expert geologists and planetary scientists, this behavior implies that Europa has a surface made almost entirely of water ice, like a giant Antarctic ice sheet girdling an entire world. And the changing appearance of the rifts and rills in this icy surface leads to a startling conclusion: The ice apparently floats on a worldwide ocean. Only by invoking liquid beneath the icy surface can scientists satisfactorily explain what they have seen, thanks to the stunning successes of the *Voyager* and *Galileo* spacecraft. Since we observe changes on the surface all around Europa, we may conclude that a worldwide ocean of liquid must underlie that surface.

What liquid could this be, and why should that substance remain liquid? Impressively, planetary scientists have reached two fairly firm additional conclusions: The liquid is water, and it remains liquid because of tidal effects on Europa produced by the giant planet Jupiter. The fact that water molecules are more abundant than ammonia, ethane, or methyl alcohol makes it the likeliest substance to provide the liquid beneath Europa's ice, and the existence of this frozen water likewise implies that more water exists in the immediate neighborhood. But how can water remain a liquid, when the solar-induced temperatures in Jupiter's vicinity are only about 120° K (−150° Celsius)? Europa's interior remains relatively warm because tidal forces from Jupiter and the two large moons nearby, Io and Ganymede, continuously flex the rocks within Europa as this moon changes its position with respect to neighboring objects. At all times, the sides of Io and Europa closest to Jupiter feel a stronger force of gravity from the giant planet than the sides farthest away. These differences in force slightly elongate the solid moons in the direction facing Jupiter. But as the moons' distances from Jupiter change during their orbits, Jupiter's tidal effect—the difference in force exerted on the near side and the far side—also changes, producing small pulses in their already distorted shapes. This changing distortion heats the moons' interiors. Like a squash ball or a racquet ball continually being smashed by impact, any system that undergoes continuing structural stress will have its internal temperature rise.

With a distance from the Sun that would otherwise guarantee a forever-frozen ice world, Io's stress level earns it the title of the most geologically active place in the entire solar system—complete with belching volcanoes, surface fissures, and plate tectonics. Some have analogized modern-day Io to the early Earth, when our planet was still piping hot from its episode of formation. Inside Io, the temperature rises to the point that volcanoes continually blast evil-smelling compounds of sulfur and sodium many miles above the satellite's surface. Io in fact has too high a temperature for liquid water to survive, but Europa, which undergoes less tidal flexing than Io because it is farther from Jupiter, heats more modestly, though still significantly. In addition, Europa's worldwide ice cap puts a pressure lid on the liquid below, preventing the water from evaporating and allowing it to exist for billions of years without freezing. So far as we can tell, Europa was born with its water ocean and ice above, and has maintained that ocean, close to the freezing point but still above it, through four and a half billion years of cosmic history.

Astrobiologists therefore view Europa's worldwide ocean as a prime target for investigation. No one knows the ice cap's thickness, which might range from a few dozen yards to half a mile or more. Given the fecundity of life within Earth's oceans, Europa remains the most tantalizing place in the solar system to search for life outside of Earth. Imagine going ice-fishing there. Indeed, engineers and scientists at the Jet Propulsion Laboratory in California have begun to envision a space probe that lands, finds (or cuts) a hole in the ice, and drops a submersible camera to have a peek at primitive life that may swim or crawl below.

"Primitive" pretty much sums up our expectations, because any would-be forms of life would have only small amounts of energy at their disposal. Nevertheless, the discovery of enormous masses of organisms at depths a mile or more beneath the basalts of Washington State, living mainly on geothermal heat, suggests that we may someday find the Europan oceans alive with organisms unlike any on Earth. But one pressing question remains: Would we call the creatures Europans or Europeans?

Mars and Europa offer targets numbers one and two in the search for extraterrestrial life within the solar system. A third great "Search Me" sign appears twice as far from the Sun as Jupiter and its moons. Saturn has one giant moon, Titan, which ties with Jupiter's champion, Ganymede, as the largest moon in the solar system. Half again as large as our own Moon, Titan possesses a thick atmosphere, a quality unequaled by any other moon (or by the planet Mercury, not much larger than Titan but much closer to the Sun, whose heat evaporates any Mercurian gases). Unlike the atmospheres of Mars and Venus, Titan's atmosphere, many dozen times thicker than Mars', consists primarily of nitrogen molecules, just as Earth's does. Floating within this transparent nitrogen gas are enormous numbers of aerosol particles, a permananet Titanian smog, that forever shrouds the moon's surface from our gaze. As a result, speculation about life's possibilities has enjoyed a field day on Titan. We have measured the moon's temperature by bouncing radio waves (which penetrate the atmospheric gases and aerosols) from its surface. Titan's surface temperature, close to 85° Kelvin (−188° Celsius), falls far below those that allow liquid water to exist, but provides just the right temperature for liquid ethane, a carbon-hydrogen compound familiar to those who refine petroleum products. For decades, astrobiologists have imagined ethane lakes on Titan, chockful of organisms that float, eat, meet, and reproduce.

Now, during the first decade of the twenty-first century, exploration has finally replaced speculation. The *Cassini-Huygens* mission to Saturn, a collaboration of NASA with the European Space Agency (ESA), left Earth in October 1997. Nearly seven years later, having received gravity boosts from Venus (twice) Earth (once), and Jupiter (once), the spacecraft reached the Saturn system, where it fired its rockets to achieve an orbit around the ringed planet.

The scientists who designed the mission arranged for the Huygens probe to detach itself from the *Cassini* spacecraft late in 2004, to make the first descent through Titan's satellite's opaque clouds, and to reach the moon's surface, using a heat shield to avoid frictional burning from its rapid passage through the upper atmosphere and a series of parachutes to slow the probe down in the lower atmosphere. Six instruments aboard the *Huygens* probe were built to measure the temperature, density, and chemical composition of Titan's atmosphere, and to send images back to Earth via the *Cassini* spacecraft. At this time, we can only await these data and images to see what they tell us about the enigma that lies beneath the clouds of Titan. We are unlikely to see life itself, should any exist on this faraway moon, but we can expect to determine whether or not conditions do favor the existence of life by providing liquid pools and ponds in which life might originate and flourish. At the very least, we may expect to learn the array of different types of molecules that exist on and near the surface of Titan, which may shed new light on how the precursors of life arose on Earth and throughout the solar system.

If we require water for life, must we restrict ourselves to planets and their moons, on whose solid surfaces water can accumulate in quantity? Not at all. Water molecules, along with several other household chemicals such as ammonia and methane and ethyl alcohol, appear routinely in cool interstellar gas clouds. Under special conditions of low temperature and high density, an ensemble of water molecules can be induced to transform and to funnel energy from a nearby star into an amplified, high-intensity beam of microwaves. The atomic physics of this phenomenon resembles what a laser does with visible light. But in this case, the relevant acronym is maser, for **m**icrowave **a**mplification by the **s**timulated **e**mission of **r**adiation. Not only does water occur practically everywhere in the galaxy, but it also occasionally beams at you as well. The great problem faced by would-be life in interstellar clouds arises not from a lack of raw materials but from the extremely low densities of matter, which enormously reduce the rate at which particles collide and interact. If life takes millions of years to arise on a planet such as Earth, it might take trillions of years to do so at much lower densities—far more time than the universe has so far provided.

By completing our search for life in the solar system, we might seem to have finished our tour through the fundamental questions linked to our cosmic origins. We cannot, however, leave this arena without a look at the great origin issue that lies in the future: the origin of our contact with other civilizations. No astronomical topic grips the public imagination more vividly, and none offers a better chance to draw together the strands of what we have learned about the universe. Now that we know something about how life might begin on other worlds, let's examine the chances of satisfying a human desire as deep as any, the wish to find other beings in the cosmos with whom we might talk things over.

CHAPTER 17

Searching for Life
in the Milky Way Galaxy

We have seen that within our own solar system, Mars, Europa, and Titan offer the best hopes for discovering extraterrestrial life, either alive or in fossil form. These three objects present by far the best chances for finding water or another substance capable of providing a liquid solvent within which molecules can meet to carry on life's work.

Because only these three objects seem likely to have pools or ponds, most astrobiologists limit their hopes of finding life in the solar system to the discovery of primitive forms of life on one or more of them. Pessimists have a reasonable argument, some day to be upheld or refuted by actual exploration, that even though we may well find conditions suitable for life on one or more of this favored threesome, life itself may well prove entirely absent. Either way, the results of our searches on Mars, Europa, and Titan will be laden with significance in judging the prevalence of life in the cosmos. Optimists and pessimists already agree on one conclusion: If we hope to find advanced forms of life—life that consists of creatures larger than the simple, single-celled organisms that appeared first and remain dominant in Earthlife—then we must look far beyond the solar system, to planets that orbit stars other than the Sun.

Once upon a time, we could only speculate about the existence of these planets. Now that well over a hundred exosolar planets have been found, basically similar to Jupiter and Saturn, we may confidently predict that only time and more precise observations separate us from the discovery of Earth-like planets. The final years of the twentieth century seem likely to mark the moment in history when we acquired real evidence for an abundance of habitable worlds throughout the cosmos. Thus the first two terms in the Drake equation, which together measure the numbers of planets orbiting stars that last for billions of years, now imply high rather than low values. The next two terms, however, which describe the probability of finding planets suitable for life, and of life actually springing into existence on such planets, remain nearly as uncertain as they did before the discovery of exosolar planets. Even so, our attempts to estimate these probabilities seem to rest on firmer grounds than our numbers for the final two terms: the probability that life on another world will evolve to produce an intelligent civilization, and the ratio of the average amount of time that such a civilization will survive to the lifetime of the Milky Way galaxy.

For the first five terms in the Drake equation, we can offer our planetary system and ourselves as a representative example, though we must always invoke the Copernican principle to avoid measuring the cosmos against ourselves, rather than the reverse. When we get to the equation's final term, however, and attempt to estimate the average lifetime of a civilization once it has acquired the technological capacity to send signals across interstellar distances, we fail to reach an answer even if we take Earth as a guide, since we have yet to determine how long our own civilization will last. We have now possessed interstellar-signaling capacity for nearly a century, ever since powerful radio transmitters began to send messages across Earth's oceans. Whether we last as a civilization for the next century, through the next millennium, or throughout a thousand centuries depends on factors far beyond our capacity to foresee, though many of the signs seem unfavorable to our long-term survival.

Asking whether our own fate corresponds to the average in the Milky Way takes us into another dimension of speculation, so the final term in the Drake equation, which affects the result as directly as all the others, may be judged just plain unknown. If, in an optimistic assessment, most planetary systems contain at least one object suitable for life, and if life originates on a sensibly high fraction (say one tenth) of those suitable objects, and if intelligent civilizations likewise appear on, perhaps, one tenth of the objects with life, then at some point in the history of the Milky Way's 100 billion stars, 1 billion locations could produce an intelligent civilization. This enormous number springs, of course, from the fact that our galaxy contains so many stars, most of them much like our Sun. For a pessimistic view of the situation, simply change each the numbers to which we assigned values from one tenth to one chance in ten thousand. Then the billion locations become 1,000, lower by a factor of 1 million.

This makes a major difference. Suppose that an average civilization, qualifying as a civilization by possessing interstellar communications ability, lasts for 10,000 years—approximately one part in a million of the Milky Way's lifetime. On the optimistic view, a billion places give birth to a civilization at some point in history, so at any representative time, about 1,000 civilizations should be flourishing. The pessimistic view, in contrast, implies that in each representative era about 0.001 civilization should exist, making ourselves a lone and lonely blip that temporarily rises high above the average value.

Which estimate has the greater chance of coming close to the true value? In science, nothing convinces so well as experimental evidence. If we hope to determine the average number of civilizations in the Milky Way, the best scientific approach would measure how many civilizations now exist. The most direct way to perform that feat would survey the entire galaxy, as the cast of television's *Star Trek* love to do, noting the number and type of each civilization that we encounter, if indeed we find any. (The possibility of an alien-free galaxy makes for boring television, rarely appearing on the small screen.) Unfortunately, this survey lies far outside our current technological capability and budgetary constraints.

Besides, surveying the entire galaxy would take millions of years, if not longer. Consider what a television program about interstellar space surveys would be like if it limited itself by what we know of physical reality. A typical hour would show the crew complaining and bickering, aware that they had come so far yet still had so far to go. "We've read all the magazines," one of them might remark. "We're sick of each other, and you, Captain, are a great pain in the plethora." Then, while other crew members sang songs to themselves and still others entered private worlds of madness, a trailing long shot would remind us that the distances to other stars in the Milky Way are millions of times greater than the distances to other planets in the solar system.

Actually, this ratio describes only the distances to the Sun's closer neighbors, already so distant that their light takes many years to reach us. A full tour of the Milky Way would take us nearly ten thousand times farther. Hollywood films depicting interstellar space flight deal with this all-important issue by ignoring it (*Invasion of the Body Snatchers, 1956* and *1978*), assuming that better rockets or improved understanding of physics will deal with it (*Star Wars*, 1977), or offering intriguing

approaches such as freeze-drying astronauts so that they can survive immensely long journeys (*Planet of the Apes*, 1968).

All of these approaches have a certain appeal, and some offer creative possibilities. We may indeed improve our rockets, which can
now reach speeds of only about one ten-thousandth of the speed of
light, the fastest we can hope to travel according to our current
knowledge of physics. Even at the speed of light, however, travel to
the nearest stars will take many years, and travel across the Milky
Way close to a thousand centuries. Freeze-drying astronauts has
some promise, but so long as those on Earth, who presumably will
pay for the trip and remain unfrozen, the long passages of time
before the astronauts return argues against easy funding. Given our
short attention spans, by far the better approach to establishing contact with extraterrestrial civilizations—provided that they exist—
appears right here on Earth. All we need do is to wait for them to
contact us. This costs far less and can offer the immediate rewards
that our society so eagerly craves.

Only one difficulty arises: Why should they? Just what about
our planet makes us special to the point that we merit attention
from extraterrestrial societies, assuming that they exist? On this
point more than any other, humans have consistently violated the
Copernican principle. Ask anyone why Earth deserves scrutiny,
and you are likely to receive a sharp, angry stare. Almost all con-
ceptions of alien visitors to Earth, as well as a sizable part of religious dogma, rest on the unspoken, obvious conclusion that our
planet and our species rank so high on the list of universal mar-
vels that no argument is needed to support the astronomically
strange contention that our speck of dust, nearly lost in its Milky
Way suburb, somehow stands out like a galactic beacon, not only
demanding but also receiving attention on a cosmic scale.

This conclusion springs from the fact that the actual situation
appears reversed when we view the cosmos from Earth. Then
planetary matters bulk large, while the stars seem tiny points of

light. From a quotidian point of view, this makes complete sense. Our success at survival and reproduction, like that of every other organism, has little to do with the cosmos that surrounds us. Among all astronomical objects, only the Sun, and to a much lesser extent the Moon, affect our lives, and their motions repeat with such regularity that they almost seem part of the Earth-bound scene. Our human consciousness, formed on Earth from countless encounters with terrestrial creatures and events, understandably renders the extraterrestrial scene as a far-distant backdrop to the important action at center stage. Our error lies in assuming that the backdrop likewise regards ourselves as the center of activity.

Because each of us adopted this erroneous attitude long before our conscious minds attained any dominion or control over our patterns of thinking, we cannot eliminate it entirely from our approach to the cosmos even when we choose to do so. Those who impose the Copernican principle must remain ever vigilant against the murmurings of our reptilian brains, assuring us that we occupy the center of the universe, which naturally directs its attention our way.

When we turn to reports of extraterrestrial visitors to Earth, we must recognize another fallacy of human thought, as omnipresent and self-deceiving as our anti-Copernican prejudices. Human beings trust their memories far more than reality can justify. We do so for the same survival value reasons that we regard Earth as the center of the cosmos. Memories record what we perceive, and we do well to pay attention to this record if we seek to draw conclusions for the future.

Now that we have better means of recording the past, however, we know better than to rely on individual memories for all matters of importance to society. We transcribe congressional debates and laws in print, videotape crime scenes, and make surreptitious audio recordings of criminal activity, because we recognize these media as superior to our own brains for creating a permanent record of past events. One great apparent exception to this rule remains. We continue to accept eyewitness testimony as accurate, or at least probative, in legal proceedings. We do so despite test after test that demonstrates that each of us, despite our best intentions, will fail to remember events accurately, especially when those memories—as they usually do in cases important enough to go to trial—deal with unusual and exciting occurrences. Our legal system accepts eyewitness testimony from long traditition, because of its emotional resonance, and most of all because it often provides the only direct evidence of past events. Nevertheless, every courtroom cry of "That's the man who held the pistol!" must be weighed against the many demonstrated cases where that was not the man, despite the witness's sincere belief to the contrary.

If we bear these facts in mind when we analyze reports of unidentified flying objects (UFOs), we can immediately recognize an enormous potential for error. By definition, UFOs are bizarre occurrences, which cause observers to discriminate among familiar and unfamiliar objects on the rarely examined celestial backdrop, and typically require rapid conclusions about these objects before they quickly disappear. Add to this the psychic charge arising from the observer's belief in having witnessed a tremendously unusual event, and we could hardly find a better textbook example of a situation likely to generate an erroneous memory.

What can we do to obtain data on UFO reports more reliable than eyewitness accounts? In the 1950s, astrophysicist J. Allen Hynek, then a leading Air Force consultant on UFOs, liked to highlight this issue by whipping a miniature camera from his pocket, insisting that if he ever saw a UFO, he would use the camera to obtain valid scientific evidence, because he knew that eyewitness testimony would not qualify. Unfortunately, improvements in technology since that time allow the creation of fake images and video recordings barely distinguishable from honest ones, so that Hynek's plan would no longer allow us to put our faith in photographic evidence supporting a UFO sighting. In fact, when we consider the interaction of memory's fragile power with the inventiveness of human con artists, we cannot easily devise a test to discriminate between fact and fancy for any individual UFO sighting.

When we turn to the more modern phenomenon of UFO abductions, the ability of the human psyche to trump reality becomes even more apparent. Although hard numbers cannot be easily obtained, in recent years tens of thousands of people have apparently come to believe that they have been taken aboard an alien spacecraft and subjected to examinations, often of the most humiliating variety. From a calm perspective, stating this claim suffices to refute it as reality. Direct application of the principle of Occam's razor, which calls for the simplest explanation that fits the alleged facts, leads to the conclusion that these abductions have been imagined, not undergone. Because nearly all of the retellings place the abduction deep in the nighttime, and the majority in the midst of sleep, the likeliest explanation involves the hypnagogic state, the boundary between sleep and waking. For many people, this state brings visual and auditory hallucinations, and sometimes a "waking dream," in which the person feels conscious but unable to move. These effects pass through the filters of our brains to yield seemingly real memories, capable of arousing unshakable belief in their certainty.

Compare this explanation of UFO abductions with an alternative, that extraterrestrial visitors have singled out Earth and arrived in sufficient numbers to abduct humans by the thousands, though only briefly, and apparently to examine them closely (but should they not have long ago learned whatever they cared to— and could they not abduct sufficient corpses to learn human anatomy in detail?). Some stories imply that aliens extract some useful substances from their abductees, or plant their seeds into female victims, or alter their mind patterns to avoid later detection (but in that case could they not eliminate abduction memories entirely?). These assertions cannot be dismissed categorically, any more than we can rule out the possibility that alien visitors wrote these words, attempting to lull human readers into a false sense of security that will further the aliens' plans for world or cosmic domination. Instead, relying upon our ability to analyze situations rationally, and to discriminate between more likely and less likely explanations, we can assign an extremely low probability to the abduction hypothesis.

One conclusion seems unassailable by UFO skeptics and believers alike. If extraterrestrial societies do visit Earth, they must know that we have created worldwide capabilities for disseminating information and entertainment, if not for distinguishing one from the other. To say that these facilities would be open to any alien visitors caring to use them amounts to a gross understatement. They would receive immediate permission (come to think of it, they might not need it), and could make their presence felt in a minute—if they cared to. The absence of apparent extraterrestrials from our television screens testifies either to their absence from Earth or to their unwillingness to reveal themselves to our gaze—the "shyness" problem. The second explanation raises an intriguing conundrum. If alien visitors to Earth choose not to be detected, and if they possess technology far superior to ours, as their journeys across interstellar distances imply, why can they simply not succeed in their plans? Why should we expect to have any evidence—visual sightings, crop circles, pyramids built by ancient astronauts, memories of abductions—if the aliens prefer that we don't? They must be messing with our minds, enjoying their little game of cat and mouse. Quite probably they are secretly manipulating our leaders too, a conclusion that snaps much of politics and entertainment into immediate focus.

The UFO phenomenon highlights an important aspect of our

consciousness. Believing though we do that our planet forms the center of creation, and that our starry surroundings must decorate our world, rather than the reverse, we nevertheless maintain a strong desire to connect with the cosmos, manifested in mental activities as disparate as credence in extraterrestrial visitor reports and belief in a benevolent deity that sends thunderbolts and emissaries to Earth. The roots of this attitude lie in the days when a self-evident distinction existed between the sky above and Earth below, between the objects we could touch and scratch and those that moved and shone but remained forever beyond our reach. From these differences we drew distinctions between the earthly body and the cosmic soul, the mundane and the marvelous, the natural and supernatural. The need for a mental bridge connecting these two apparent aspects of reality has informed many of our attempts to create a coherent picture of our existence. Modern science's demonstration that we are stardust has thrown an enormous wrench into our mental equipment, from which we are still struggling to recover. UFOs suggest new messengers from the other part of existence, all-powerful visitors who well know what they are up to while we remain ignorant, barely aware that the truth is out there. This attitude was captured well in the classic film *The Day the Earth Stood Still* (1951), in which an alien visitor, far wiser than we, comes to Earth to warn that our violent behavior may lead to our own destruction.

Our innate feelings about the cosmos manifest a dark side that projects our feelings about human strangers onto nonhuman visitors. Many a UFO report contains phrases similar to "I heard something odd outside, so I took my rifle and went to see what it was." Films that depict aliens on Earth likewise slip easily into a hostile mode, from the cold war epic *Earth Versus the Flying Saucers* (1956), in which the military blasts away at alien spacecraft without pausing to ask their intentions, to *Signs* (2002), in which the peace-loving hero, with no rifle at hand, uses a baseball

bat to chastise his trespassers—a method not likely to succeed
against actual aliens capable of crossing interstellar distances.

The greatest arguments against interpreting UFO reports as
evidence for extraterrestrial visitors reside in our planet's unimportance, together with the vast distances between the stars. Neither can be regarded as absolute bars to this interpretation, but in
tandem they form a powerful argument. Must we, then, conclude
that because Earth lacks popular appeal, our hopes of finding
other civilizations must await the day when we can expend our
own resources to embark on journeys to other planetary systems?

Not at all. The scientific approach to establishing contact with
other civilizations within the Milky Way and beyond, should they
exist, has always relied on letting nature work in our favor. This
principle redirects the question: What aspect of extraterrestrial
civilizations would we find most exciting? (answer: Visitors in the
flesh) into the scientifically fruitful one: What seems to be the
most likely means of establishing contact with other civilizations?
Nature, and the immense distances between stars, supply the
answer—use the cheapest, fastest means of communication ava-
iable, which presumably holds the same rank elsewhere in the
galaxy.

The cheapest and fastest way to send messages between the stars
uses electromagnetic radiation, the same medium that carries
almost all long-range communication on Earth. Radio waves have
revolutionized human society by allowing us to send words and
pictures around the world at 186,000 miles per second. These messages travel so rapidly that even if we beam them up to a stationary satellite orbiting at an altitude of 23,000 miles, which relays
them to another part of Earth's surface, they undergo a time delay
on each leg of their journey much shorter than one second.

Over interstellar distances, the time lag grows longer, though it
remains the shortest we can hope to achieve. If we plan to send a
radio message to Alpha Centauri, the star system closest to the

Sun, we must plan on a travel time of 4.4 years in each direction. Messages that travel for, say, twenty years can reach several hundred stars, or any planets that orbit them. Thus if we are prepared to wait for a round-trip of forty years, we could beam a message toward each of these stars, and eventually find out whether we receive a reply from any of them. This approach assumes, of course, that if civilizations exist close to any of these stars, they have a command of radio, and an interest in its application, at least equal to ours.

The fundamental reason why we don't adopt this approach in searching for other civilizations lies not in its assumptions but in our attitudes. Forty years is a long time to wait for something that may never happen. (Yet if we had beamed out messages forty years ago, by now we would have some serious information about the abundance of radio-using civilizations in our region of the Milky Way.) The only serious attempt in this direction occurred in the 1970s, when astronomers celebrated the upgrading of the radio telescope near Arecibo, Puerto Rico, by using it to beam a message for a few minutes in the direction of the star cluster M13. Since the cluster lies 25,000 light-years away, any return message will be a long time coming, rendering the exercise more a demonstration than an actual casting call. In case you think that discretion has inhibited our broadcasting (for it is good to be shifty in a new country), recall that all of our post–World War II radio and television broadcasting, as well as our powerful radar beams, have sent spherical shells of radio waves into space. Expanding at the speed of light, the "messages" from the *Honeymooners* and *I Love Lucy* era have already washed over thousands of stars, while *Hawaii Five-O* and *Charlie's Angels* have reached hundreds. If other civilizations really could disentangle individual programs from the cacaphony of Earth's radio emission—now comparable to or stronger than that from any solar system object, including the Sun—there might be some truth to the playful speculation that the content of these programs explains why we have heard nothing from our neighbors, because they find our programming either so appalling or (dare we suggest) so overwhelmingly impressive that they choose not to reply.

A message might arrive tomorrow, laden with intriguing information and commentary. Herein lies the greatest appeal of communication by electromagnetic radiation. Not only is it cheap (sending fifty years of television broadcasts into space has cost lest than a single spacecraft mission), it is also instantaneous—provided that we can receive and interpret another civilization's emission. This also provides a fundamental aspect of UFO excitement, but in this case we might actually receive transmissions that could be recorded, verified as real, and studied for as long as it would take to understand them.

In the search for extraterrestrial intelligence, shortened to SETI by the scientists who engage in it, the focus remains on searching for radio signals, though the alternative of looking for signals sent with light waves should not be rejected. Although light waves from another civilization must compete with myriad natural sources of light, laser beams offer the opportunity to concentrate the light into a single color or frequency—the same approach that allows radio waves to carry messages from different radio or television stations. So far as radio waves go, our hopes for success in SETI rest with antennas that can survey the sky, receivers that record what the antennas detect, and powerful computers that analyze the receivers' signals in a search for the unnatural. Two basic possibilities exist: We might find another civilization by eavesdropping on its own communications, some of which leaks into space in the same way that our radio and television broadcasts do; or we might discover deliberately beamed signals, meant to attract the attention of previously uncatalogued civilizations such as our own.

Eavesdropping clearly presents a more difficult task. A beamed signal concentrates its power in a particular direction, so that detecting that signal becomes much easier if it should be deliberately sent toward us, whereas signals that leak into space diffuse their power more or less evenly in all directions and are therefore much weaker at a particular distance from their source than a beamed signal. Furthermore, a beamed signal would presumably contain some easy warm-up exercises to tell its recipients how to interpret it, whereas radiation that leaks into space presumably carries no such user's manual. Our own civilization has leaked signals for many decades, and has sent a beamed signal in one particular direction for a few minutes. If civilizations are rare, any attempts to find them ought to concentrate on eavesdropping and avoid the lure of hoping for deliberately beamed signals.

With ever better systems of antennas and receivers, SETI proponents have begun to eavesdrop on the cosmos, hoping to find evidence for other civilizations. Precisely because we have no guarantee that we shall ever hear anything by eavesdropping, those who engage in these activities have had difficulty securing funding. In the early 1990s, the U.S. Congress supported a SETI program for a year, until cooler heads pulled the plug. SETI scientists now draw their supports, in part, from millions of people who download a screen saver (from the Web site setiathome.sl .berkeley.edu) that co-opts home computers to analyze data for alien signals in their spare time. Even more funding has come from wealthy individuals, most notably the late Bernard Oliver, a prominent Hewlett-Packard engineer with a lifelong interest in SETI, and Paul Allen, the co-founder of Microsoft. Oliver spent many years thinking about the basic problem in SETI, the difficulty of searching through billions of possible frequencies at which other civilizations might be broadcasting. We divide the radio spectrum into relatively wide bands, so that only a few hundred different frequencies exist for radio and television broadcasts. In principle, however, alien signals might be confined so narrowly in frequency that the SETI dial would need billions of entries. Powerful computer systems, which lie at the heart of current SETI efforts, can meet this challenge by analyzing hundreds of millions of frequencies simultaneously. On the other hand, they have not yet found anything suggestive of another civilization's radio communications.

More than fifty years ago, the Italian genius Enrico Fermi, perhaps the last great physicist to work both as an experimentalist and as a theorist, discussed extraterrestrial life during lunch with his colleagues. Agreeing that nothing particularly special distinguishes Earth as an abode for life, the scientists reached the conclusion that life ought to be abundant in the Milky Way. In that case, Fermi asked, in a query that ripples across the decades, *where are they?*

Fermi meant that if many places in our galaxy have seen the advent of technologically advanced civilizations, surely we should have heard from one of them by now, by radio or laser messages if not by actual visits. Even if most civilizations die out quickly, as ours may, the existence of large numbers of civilizations implies that some of them should have sufficiently extended lifetimes to mount long-term searches for others. Even if some of these long-lived civilizations do not care to engage in such searches, others will. So the fact that we have no scientifically verified visits to Earth, nor reliable demonstrations of signals produced by another civilization, may prove that we have badly overestimated the likelihood that intelligent civilizations arise in the Milky Way.

Fermi had a point. Every day that passes adds a bit more evidence that we may be alone in our galaxy. However, when we examine the actual numbers, the evidence looks weak. If several thousand civilizations exist in the galaxy at any representative time, the average separation between neighboring civilizations will be a few thousand light-years, a thousand times the distance to the closest stars. If one or more of these civilizations has lasted for millions of years, we might expect that by now they should have sent us a signal, or revealed themselves to our modest eavesdropping efforts. If, however, no civilization attains anything like this age, then we shall have to work harder to find our neighbors, because none of them may be engaged in a galaxywide attempt to find others, and none of them may be broadcasting so powerfully that our present eavesdropping efforts can find them.

Thus we remain in a familiar human condition, poised at the edge of events that may not occur. The most important news in human history could arrive tomorrow, next year, or never. Let us go forth into a new dawn, ready to embrace the cosmos as it surrounds us, and as it reveals itself, shining with energy and replete with mystery.

The Search for Ourselves in the Cosmos

Equipped with his five senses, man explores the universe around him and calls the adventure science.
—Edwin P. Hubble, 1948

Human senses display an astonishing acuity and range of sensitivity. Our ears can record the thunderous launch of the space shuttle, yet they can also hear a male mosquito buzzing in the corner of a room. Our sense of touch allows us to feel the crush of a bowling ball dropped on our big toe, or to tell when a one-milligram bug crawls along our arm. Some people enjoy munching on habanero peppers, while sensitive tongues can identify the presence of food flavors at a few parts per million. And our eyes can register the bright sandy terrain on a sunny beach, yet have no trouble spotting a lone match, freshly lit hundreds of feet away, across a darkened auditorium. Our eyes also allow us to see across the room and across the universe. Without our vision, the science of astronomy would never have been born and our capacity to measure our place in the universe would have remained hopelessly stunted.

In combination, these senses allow us to decode the basics of our immediate environment, such as whether it's day or night, or when a creature is about to eat you. But little did anybody know, until the last few centuries, that our senses alone offer only a narrow window on the physical universe.

Some people boast of a sixth sense, professing to know or see things that others cannot. Fortunetellers, mind readers, and mystics top the list of those who claim mysterious powers. In doing so, they instill widespread fascination in others. The questionable field of parapsychology rests on the expectation that at least some people actually harbor this talent.

In contrast, modern science wields dozens of senses. But scientists do not claim that these are the expression of special powers, just special hardware that converts the information gleaned by these extra senses into simple tables, charts, diagrams, or images that our five inborn senses can interpret.

With apologies to Edwin P. Hubble, his remark quoted on page 291, while poignant and poetic, should instead have been

> Equipped with our five senses, along with telescopes and microscopes and mass spectrometers and seismographs and magnetometers and particle detectors and accelerators and instruments that record radiation from the entire electromagnetic spectrum, we explore the universe around us and call the adventure science.

Think of how much richer the world would appear to us, and how much sooner we would have discovered the fundamental nature of the universe, if we were born with high-precision, tunable eyeballs. Dial up the radio-wave part of the spectrum and the daytime sky turns as dark as night, except for some choice directions. Our galaxy's center appears as one of the brightest spots on the sky, shining brightly behind some of the principal stars of the constellation Sagittarius. Tune into microwaves and the entire universe glows with a remnant from the early universe, a wall of light that set forth on its journey to us 380,000 years after the big bang. Tune into X rays and you will immediately spot the locations of black holes with matter spiraling into them. Tune into gamma rays and see titanic explosions bursting forth from random directions about once a day throughout the universe. Watch the effect of these explosions on the surrounding material as it heats up to produce X rays, infrared, and visible light.

If we were born with magnetic detectors, the compass would never have been invented because no one would ever need one. Just tune into Earth's magnetic field lines and the direction of magnetic North looms like Oz beyond the horizon. If we had spectrum analyzers within our retinas, we would not have to wonder what the atmosphere is made of. Simply by looking at it we would know whether or not it contains sufficient oxygen to sustain human life. And we would have learned thousands of years ago that the stars and nebulae in our galaxy contain the same chemical elements as those found here on Earth.

And if we were born with big, sensitive eyes and built-in Doppler motion detectors, we would have seen immediately, even as grunting troglodytes, that the entire universe is expanding— that all distant galaxies are receding from us.

If our eyes had the resolution of high-performance microscopes, nobody would have ever blamed the plague and other sicknesses on divine wrath. The bacteria and viruses that made you sick would have been in plain view as they crawled on your food or slid through open wounds in your skin. With simple experiments, you could easily tell which of these bugs were bad and which were good. And the carriers of postoperative infection problems would have been identified and solved hundreds of years earlier.

If we could detect high-energy particles, we would spot radioactive substances from great distances. No Geiger counters necessary. You could even watch radon gas seep through the basement floor of your home and not have to pay somebody to tell you about it.

The honing of our five senses from birth through childhood allows us as adults to pass judgment on events and phenomena in our lives, declaring whether or not they "make sense." Problem is, hardly any scientific discoveries of the past century have flowed from the direct application of our senses. They came instead from the direct application of sense-transcendent mathematics and hardware. This simple fact explains why, to the average person, relativity, particle physics, and eleven-dimensional string theory make no sense. Add to this list black holes, wormholes, and the big bang. Actually, these concepts don't make much sense to scientists either, until we have explored the universe for a long time with all senses that are technologically available. What eventually emerges is a newer and higher level of "uncommon sense" that enables scientists to think creatively and to pass judgment in the unfamiliar underworld of the atom or in the mind-bending domain of higher dimensional space. The twentieth-century German physicist Max Planck made a similar observation about discovery of quantum mechanics: "Modern physics impresses us particularly with the truth of the old doctrine which teaches that there are realities existing apart from our sense-perceptions, and that there are problems and conflicts where these realities are of greater value for us than the richest treasures of the world of experience."

Each new way of knowing heralds a new window on the universe—a new detector to add to our growing list of nonbiological senses. Whenever this happens, we achieve a new level of cosmic enlightenment, as though we were evolving into supersentient beings. Who could have imagined that our quest to decode the mysteries of the universe, armed with a myriad of artificial senses, would grant us insight into ourselves? We embark on this quest not from a simple desire but from a mandate of our species to search for our place in the cosmos. The quest is old, not new, and has garnered the attention of thinkers great and small, across time and across culture. What we have discovered, the poets have known all along:

We shall not cease from exploration
And the end of all our exploring
Will be to arrive where we started
And know the place for the first time . . .

—T. S. Eliot, 1942

Glossary of Selected Terms

absolute (Kelvin) temperature scale: *Temperature* measured on a scale (denoted by K) on which water freezes at 273.16 K and boils at 373.16 K, with 0 K denoting absolute zero, the coldest theoretically attainable temperature.

acceleration: A change in an object's speed or direction of motion (or both).

accretion: An infall of matter that adds to the mass of an object.

accretion disk: Material surrounding a massive object, typically a *black hole*, that moves in orbit around it and slowly spirals inward.

AGN: Astronomical shorthand for a *galaxy* with an active *nucleus*, a modest way of describing galaxies whose central regions shine thousands, millions, or even billions of times more brightly than the central regions of a normal galaxy. AGNs have a generic similarity to *quasars*, but they are typically observed at distances less than that of quasars, hence later in their lives than quasars themselves.

amino acid: One of a class of relatively small *molecules*, made of thirteen to twenty-seven *atoms* of *carbon*, *nitrogen*, *hydrogen*, *oxygen*, and *sulfur*, which can link together in long chains to form *protein molecules*.

Andromeda galaxy: The closest large *spiral galaxy* to the Milky Way, approximately 2.4 million *light-years* from our own galaxy.

antimatter: The complementary form of matter, made of *antiparticles* that have the same mass but opposite sign of *electric charge* as the particles that they complement.

antiparticle: The *antimatter* complement to a particle of ordinary matter.

apparent brightness: The brightness that an object appears to have as an observer measures it, hence a brightness that depends on the object's *luminosity* and its distance from the observer.

Archaea: Representatives of one of the three domains of life, thought to be the oldest forms of life on Earth. All Archaea are single-celled and thermophilic (capable of thriving at temperatures above 50–70° Celsius).

asteroid: One of the objects, made primarily of rock or of rock and metal, that orbit the Sun, mainly between the orbits of Mars and Jupiter, and range in size from 1,000 kilometers in diameter down to objects about 100 meters across. Objects similar to asteroids but smaller in size are called *meteoroids*.

astronomer: One who studies the *universe*. Used more commonly in the past, at a time before spectra were obtained of cosmic objects.

astrophysicist: One who studies the *universe* using the full toolkit enabled by the known laws of physics. The preferred term in modern times.

atom: The smallest electrically neutral unit of an *element*, consisting of a *nucleus* made of one or more *protons* and zero or more *neutrons*, around which orbit a number of *electrons* equal to the number of protons in the nucleus. This number determines the chemical characteristics of the atom.

Bacteria: One of the three domains of life on Earth (formerly known as prokaryotes), single-celled organisms with no well-defined *nucleus* that holds genetic material.

barred spiral galaxy: A *spiral galaxy* in which the distribution of stars and gas in the galaxy's central regions has an elongated, barlike configuration.

big bang: The scientific description of the origin of the *universe*, premised on the hypothesis that the universe began in an explosion that brought space and matter into existence approximately 14 billion years ago. Today the universe continues to expand in all directions, everywhere, as the result of this explosion.

black hole: An object with such enormous *gravitational force* that nothing, not even light, can escape from within a specific distance from its center, called the object's *black hole radius*.

black hole radius: For any object with a mass M, measured in units of the Sun's mass, a distance equal to 3M kilometers, also called the object's *event horizon*.

blue shift: A shift to higher *frequencies* and shorter *wavelengths*, typically caused by the *Doppler effect*.

brown dwarf: An object with a composition similar to a star's, but with too little mass to become a star by initiating *nuclear fusion* in its core.

carbohydrate: A *molecule* made only of *carbon, hydrogen*, and *oxygen atoms*, typically with twice as many hydrogen as oxygen atoms.

carbon: The element that consists of *atoms* whose *nuclei* each have six *protons*, and whose different *isotopes* each have six, seven, or eight *neutrons*.

carbon dioxide: Molecules of CO_2, which each have one *carbon atom* and two *oxygen* atoms.

Cassini-Huygens **spacecraft**: The spacecraft launched from Earth in 1997 that reached Saturn in July 2004, after which the *Cassini* orbiter surveyed Saturn and its moons and released the *Huygens* probe to descend to the surface of Titan, Saturn's largest satellite.

Celsius or **Centigrade temperature scale:** The *temperature* scale named for the Swedish astronomer Anders Celsius (1701–1744), who introduced it in 1742, according to which water freezes at zero degrees and boils at 100 degrees.

carbon dioxide (CO_2): A type of *molecule* containing one *carbon* and two *oxygen atoms*.

catalyst: A substance that increases the rate at which specific reactions between *atoms* or *molecules* occur, without itself being consumed in these reactions.

CBR: See **cosmic background radiation**.

cell: A structural and functional unit found in all forms of life on Earth.

chromosome: A single *DNA* molecule, together with the *proteins* associated with that molecule, which stores genetic information in subunits called *genes* and can transmit that information when cells *replicate*.

civilization: For SETI activities, a group of beings with interstellar communications ability at least equal to our own on Earth.

COBE (COsmic Background Explorer) satellite: The satellite launched in 1989 that observed the *cosmic background radiation* and made the first detection of small differences in the amount of this radiation arriving from different directions on the sky.

comet: A fragment of primitive solar system material, typically a "dirty snowball" made of ice, rock, dust, and frozen *carbon dioxide* (dry ice).

compound: A synonym for *molecule*.

constellation: A localized group of stars, as seen from Earth, named after an animal, planet, scientific instrument, or mythological character, which in rare cases actually describes the star pattern; one of eighty-eight such groups in the sky.

cosmic background radiation (CBR): The sea of *photons* produced everywhere in the *universe* soon after the *big bang*, which still fills the universe and is now characterized by a *temperature* of 2.73 K.

cosmological constant: The constant introduced by Albert Einstein into his equation describing the overall behavior of the *universe*, which describes the amount of energy, now called *dark energy*, in every cubic centimeter of seemingly empty space.

cosmologist: An *astrophysicist* who specializes in the origin and large-scale structure of the *universe*.

cosmology: The study of the *universe* as a whole, and of its structure and evolution.

cosmos: Everything that exists; a synonym for *universe*.

dark energy: *Energy* that is invisible and undetectable by any direct measurement, whose amount depends on the size of the *cosmological constant*, and which tends to make space expand.

dark matter: Matter of unknown form that emits no *electromagnetic radiation*, that has been deduced, from the *gravitational forces* it exerts on visible matter, to comprise the bulk of all matter in the universe.

decoupling: The era in the *universe*'s history when *photons* first had too little energy to interact with *atoms*, so that for the first time atoms could form and endure without being broken apart by photon impacts.

DNA (deoxyribonucleic acid) molecule: A long, complex *molecule* consisting of two interlinking spiral strands, bound together by thousands of cross links formed from small molecules. When DNA molecules divide and *replicate*, they split lengthwise, splitting each pair of small molecules that form their cross links. Each half of the molecule then forms a new replica of the original molecule from smaller molecules that exist in the nearby environment.

Doppler effect: The change in *frequency, wavelength,* and *energy* observed for *photons* arriving from a source that has a relative velocity of approach or recession along an observer's line of sight to the source. These changes in frequency and wavelength are a general phenomenon that occurs with any type of wave motion. They do not depend on whether the source is moving or the observer is moving; what counts is the relative motion of the source with respect to the observer along the observer's line of sight.

Doppler shift: The fractional change in the *frequency, wavelength,* and *energy* produced by the *Doppler effect.*

double helix: The basic structural shape of *DNA molecules.*

Drake equation: The equation, first derived by the American astronomer Frank Drake, that summarizes our estimate of the number of *civilizations* with interstellar communications capability that exist now or at any representative time.

dry ice: Frozen *carbon dioxide* (CO_2).

dust cloud: Gas clouds in interstellar space that are cool enough for *atoms* to combine to form *molecules,* many of which themselves combine to form dust particles made of millions of atoms each.

dynamics: The study of the motion and the effect of *forces* on the interaction of objects. When applied to the motion of objects in the solar system and the universe, this is often called celestial mechanics.

eavesdropping: The technique of attempting to detect an extraterrestrial *civilization* by capturing some of the *radio* signals used for the civilization's internal communications.

eccentricity: A measure of the flatness of an *ellipse,* equal to the ratio of the distance between the two "foci" of the ellipse to its long axis.

eclipse: The partial or total obscuration of one celestial object by another, as seen by an observer when the objects appear almost or exactly behind each other.

electric charge: An intrinsic property of *elementary particles*, which may be positive, zero, or negative; unlike signs of *electric charge* attract one another and like signs of electric charge repel one another through *electromagnetic forces*.

electromagnetic force: One of the four basic types of *forces*, acting between particles with *electric charge*, and diminishing in proportion to the square of the distance between the particles. Recent investigations have shown that these forces and *weak forces* are different aspects of a single *electro-weak force*.

electromagnetic radiation: Streams of *photons* that carry energy away from a source of photons.

electron: An *elementary particle* with one unit of negative *electric charge*, which in an *atom* orbits the atomic *nucleus*.

electro-weak forces: The unified aspect of *electromagnetic forces* and *weak forces*, whose aspects appear quite different at relatively low energies but become unified when acting at enormous energies such as those typical of the earliest moments of the *universe*.

elements: The basic components of matter, classified by the number of *protons* in the *nucleus*. All ordinary matter in the *universe* is composed of ninety-two elements that range from the smallest *atom*, *hydrogen* (with one proton in its nucleus), to the largest naturally occurring element, uranium (with ninety-two protons in its nucleus). Elements heavier than uranium have been produced in laboratories.

elementary particle: A fundamental particle of nature, normally indivisible into other particles. *Protons* and *neutrons* are usually designated as elementary particles although they each consist of three particles called *quarks*.

ellipse: A closed curve defined by the fact that the sum of the distances from any point on the curve to two interior fixed points, called foci, has the same value.

elliptical galaxy: A galaxy with an ellipsoidal distribution of stars, containing almost no interstellar gas or dust, whose shape seems elliptical in a two-dimensional projection.

energy: The capacity to do work; in physics, "work" is specified by a given amount of *force* acting through a specific distance.

energy of mass: The *energy* equivalent of a specific amount of mass, equal to the mass times the square of the speed of light.

energy of motion: See *kinetic energy*.

enzyme: A type of *molecule*, either a *protein* or *RNA*, that serves as a site at which molecules can interact in certain specific ways, and thus acts as a *catalyst*, increasing the rate at which particular molecular reactions occur.

escape velocity: For a projectile or spacecraft, the minimum speed required for an outbound object to leave its point of launching and never return to the object, despite the object's *gravitational force*.

Eukarya: The totality of organisms classified as *eukaryotes*.

eukaryote: An organism, either single-celled or multicellular, that keeps the genetic material in each of its cells within a membrane-bounded nucleus.

Europa: One of Jupiter's four large satellites, notable for its icy surface that may cover a worldwide ocean.

event horizon: The poetic name given to an object's *black hole radius*: the distance from a *black hole*'s center that marks the point of no return, because nothing can escape from the black hole's *gravitational force* after passing inward through the event horizon. The event horizon may be considered to be the "edge" of a black hole.

evolution: In biology, the ongoing result of *natural selection*, which under certain circumstances causes groups of similar organisms, called species, to change over time so that their descendants differ significantly in structure and appearance; in general, any gradual change of an object into another form or state of development.

exosolar (also **extrasolar**): Pertaining to objects beyond the *solar system*. We prefer "exo" for its correspondence with exobiology, the study of life forms with origins beyond Earth.

exosolar planet (also **extrasolar planet**): A *planet* that orbits a *star* other than the Sun.

extremophile: Organisms that thrive at high *temperatures*, typically between 70 and 100 degrees Celsius.

Fahrenheit temperature scale: The *temperature* scale named for the German-born physicist Gabriel Daniel Fahrenheit (1686–1736), who introduced it in 1724, according to which water freezes at 32 degrees and boils at 212 degrees.

fission: The splitting of a larger atomic *nucleus* into two or more smaller nuclei. The fission of nuclei larger than iron releases energy. This fission (also called atomic fission) is the source of energy in all present-day nuclear power plants.

force: The capacity to do work or to produce a physical change; an influence that tends to *accelerate* an object in the direction that the force is applied to the object.

fossil: A remnant or trace of an ancient organism.

frequency: Of *photons*, the number of oscillations or vibrations per second.

fusion: The combining of smaller *nuclei* to form larger ones. When nuclei smaller than iron fuse, energy is released. Fusion provides the primary energy source for the world's nuclear weapons, and for all stars in the universe. Also called *nuclear fusion* and *thermonuclear fusion*.

galaxy: A large group of stars, numbering from several million up to many hundred billion, held together by the stars' mutual gravitational attraction, and also usually containing significant amounts of gas and dust.

galaxy cluster: A large group of *galaxies*, usually accompanied by gas and dust and by a much greater amount of *dark matter*, held together by the mutual gravitational attraction of the material forming the galaxy cluster.

***Galileo* spacecraft**: The spacecraft sent by NASA to Jupiter in 1990, which arrived in December 1995, dropped a probe into Jupiter's atmosphere, and spent the next few years in orbit around the giant planet, photographing the planet and its large satellites.

gamma rays: The highest-*energy*, highest-*frequency*, and shortest-*wavelength* type of *electromagnetic radiation*.

gene: A section of a *chromosome* that specifies, by means of the genetic code, the formation of a specific chain of *amino acids*.

genetic code: The set of "letters" in *DNA* or *RNA* molecules, each of which specifies a particular *amino acid* and consists of three successive molecules like those that form the cross-links between the twin spirals of DNA molecules.

genome: The total complement of an organism's *genes*.

general theory of relativity: Introduced in 1915 by Albert Einstein, forming the natural extension of *special relativity theory* into the domain of *accelerating* objects, this is a modern theory of gravity that successfully explains many experimental results not otherwise explainable in terms of Newton's theory of gravity. Its basic premise is the "equivalence principle," according to which a person in a spaceship, for example, cannot distinguish whether the spaceship is accelerating through space, or whether it is stationary in a gravitational field that would produce the same acceleration. From this simple yet profound principle emerges a completely reworked understanding of the nature of gravity. According to Einstein, gravity is not a *force* in the traditional meaning of the word. Gravity is the curvature of space in the vicinity of a mass. The motion of a nearby object is completely determined by its velocity and the amount of curvature that is present. As counterintuitive as this sounds, general relativity theory explains all known behavior of gravitational systems ever studied and it predicts a myriad of even more counterintuitive phenomena that are continually verified by controlled experiment. For example, Einstein predicted that a strong gravity field should warp space and noticeably bend light in its vicinity. It was later shown that starlight passing near the edge of the Sun (as seen during a total solar eclipse) is found to be displaced from its expected position by an amount precisely matching Einstein's predictions. Perhaps the grandest application of the general theory of relativity involves the description of our expanding universe where all of space is curved from the collected gravity of hundreds of billions of galaxies. An important and currently unverified prediction is the existence of "gravitons"—particles that carry gravitational forces and communicate abrupt changes in a gravitational field like those expected to arise from a supernova explosion.

giant planet: A planet similar in size and composition to Jupiter, Saturn, Uranus, or Neptune, consisting of a solid core of rock and ice surrounded by thick layers of mainly *hydrogen* and *helium* gas, with a mass ranging from a dozen or so Earth masses up to many hundred times the mass of Earth.

gravitational forces: One of the four basic types of *forces*, always attractive, whose strength between any two objects varies in proportion to the product of the objects' masses, divided by the square of the distance between their centers.

gravitational lens: An object that exerts sufficient *gravitational force* on passing light rays to bend them, often focusing them to produce a brighter image than an observer would see without the gravitational lens.

gravitational radiation (gravity waves): *Radiation*, quite unlike *electromagnetic radiation* except for traveling at the speed of light, produced in relatively large amounts when massive objects move past one another at high speeds.

greenhouse effect: The trapping of *infrared* radiation by a planet's atmosphere, which raises the temperature on and immediately above the planet's surface.

habitable zone: The region surrounding a star within which the star's heat can maintain one or more *solvents* in a liquid state, hence a spherical shell around the star with an inner and an outer boundary.

halo: The outermost regions of a galaxy—occupying a volume much larger than the visible galaxy does—within which most of a galaxy's *dark matter* resides.

helium: The second lightest and second most abundant *element*, whose nuclei all contain two *protons* and either one or two *neutrons*. Stars generate energy through the *fusion* of *hydrogen* nuclei (*protons*) into helium nuclei.

hertz: A unit of *frequency*, corresponding to one vibration per second.

Hubble's constant: The constant that appears in *Hubble's law* and relates galaxies' distances to their recession velocities.

Hubble's law: The summary of the *universe*'s expansion as observed today, which states that the recession velocities of faraway galaxies equals a constant times the galaxies' distances from the Milky Way.

Hubble Space Telescope: The space-borne telescope launched in 1991 that has secured marvelous *visible light* images of a host of astronomical objects, owing to the fact that the telescope can observe the cosmos free from the blurring and absorbing effects inevitably produced by Earth's atmosphere.

hydrogen: The lightest and most abundant *element*, whose *nuclei* each contain one *proton* and a number of *neutrons* equal to zero, one, or two.

infrared: *Electromagnetic radiation* consisting of *photons* whose *wavelengths* are all somewhat longer, and whose *frequencies* are all somewhat higher, than those of the photons that form visible light.

initial singularity: The moment at which the expansion of the *universe* began, also called the *big bang*.

inner planets: The Sun's planets Mercury, Venus, Earth, and Mars, all of which are small, dense, and rocky in comparison to the *giant planets*.

interstellar cloud: A region of interstellar space considerably denser than average, typically spanning a diameter of several dozen *light-years*, with densities of matter that range from ten atoms per cubic centimeter up to millions of molecules per cubic centimeter.

interstellar dust: Dust particles, each made of a million or so *atoms*, probably ejected into interstellar space from the atmospheres of highly rarefied *red-giant stars*.

interstellar gas: Gas within a *galaxy* not part of any stars.

ion: An *atom* that has lost one or more of its *electrons*.

ionization: The process of converting an *atom* into an *ion* by stripping the atom of one or more *electrons*.

irregular galaxy: A *galaxy* whose shape is irregular, that is, neither *spiral* (disklike) nor *elliptical*.

isotope: *Nuclei* of a specific *element*, all of which contain the same number of *protons* but different numbers of *neutrons*.

JWST (James Webb Space Telescope): The space-borne telescope, planned to begin operations during the 2010 decade, that will supersede the *Hubble Space Telescope*, carrying a larger mirror and more advanced instruments into space.

Kelvin (absolute) temperature scale: The *temperature* scale named for Lord Kelvin (William Thomson, 1824–1907) and created during the mid-nineteenth century, for which the coldest possible temperature is, by definition, zero degrees. The temperature intervals on this scale (denoted by K) are the same as those on the *Celsius (Centigrade) temperature scale*, so that on the Kelvin scale, water freezes at 273.16 degrees and boils at 373.16 degrees.

kilogram: A unit of mass in the metric system, consisting of 1,000 grams.

kilohertz: A unit of *frequency* that describes 1,000 vibrations or oscillations per second.

kilometer: A unit of length in the metric system, equal to 1,000 meters and approximately 0.62 mile.

kinetic energy: The *energy* that an object possesses by virtue of its motion, defined as one half of the object's mass times the square of the object's speed. Thus a more massive object, such as a truck, has more kinetic energy than a less massive object, such as a tricycle, that moves at the same speed.

Kuiper Belt: The material in orbit around the Sun at distances extending from about 40 A.U. (Pluto's average distance) out to several hundred A.U., almost all of which is debris left over from the Sun's *protoplanetary disk*. Pluto is one of the largest objects in the Kuiper Belt.

Large Magellanic Cloud: The larger of the two irregular satellite *galaxies* of the *Milky Way*.

latitude: On Earth, the coordinate that measures north and south by specifying the number of degrees from the Equator (zero degrees) toward the North Pole (90° north) or the South Pole (90° south).

life: A property of matter characterized by the abilities to reproduce and to *evolve*.

light (visible light): *Electromagnetic radiation* that consists of photons whose *frequencies* and *wavelengths* fall within the band denoted as visible light, between *infrared* and *ultraviolet*.

light-year: The distance that light or other forms of *electromagnetic radiation* travel in one year, equal to approximately 10 trillion kilometers or 6 trillion miles.

Local Group: The name given to the two dozen or so *galaxies* in the immediate vicinity of the *Milky Way* galaxy. The Local Group includes the *Large and Small Magellanic Clouds* and the *Andromeda galaxy*.

logarithmic scale: A method for plotting data whereby tremendous ranges of numbers can fit on the same piece of paper. In official terms, the logarithmic scale increases exponentially (e.g., 1, 10, 100, 1,000, 10,000) rather than arithmetically (e.g., 1, 2, 3, 4, 5).

longitude: On Earth, the coordinate that measures east or west by specifying the number of degrees from the arbitrarily defined "prime meridian," the north-south line passing through Greenwich, England. Longitudes range from zero to 180 degrees east or 180 degrees west of Greenwich, thus including the 360 degrees that span Earth's surface.

luminosity: The total amount of *energy* emitted each second by an object in all types of *electromagnetic radiation.*

mass: A measure of an object's material content, not to be confused with weight, which measures the amount of *gravitational force* on an object. For objects at Earth's surface, however, mass and weight vary in direct proportion.

mass extinction: An event in the history of life on Earth, in some cases as the result of a massive impact, during which a significant fraction of all species of organisms become extinct within a geologically short interval of time.

megahertz: A unit of *frequency*, equal to 1 million vibrations or oscillations per second.

metabolism: The totality of an organism's chemical processes, measured by the rate at which the organism uses *energy*. A high-metabolism animal must consume energy (food) much more frequently to sustain itself.

meteor: A luminous streak of light produced by the heating of a *meteoroid* as it passes through Earth's atmosphere.

meteor shower: A large number of *meteors* observed to radiate from a specific point on the sky, the result of Earth's crossing the orbits of a large number of *meteoroids* within a short time.

meteorite: A *meteoroid* that survives its passage through Earth's atmosphere.

meteoroid: An object of rock or metal, or a metal-rock mixture, smaller than an *asteroid*, moving in an orbit around the Sun, part of the debris left over from the formation of the solar system or from collisions between solar-system objects.

meter: The fundamental unit of length in the metric system, equal to approximately 39.37 inches.

Milky Way: The *galaxy* that contains the Sun and approximately 300 billion other stars, as well as interstellar gas and dust and a huge amount of dark matter.

model: A mental construct, often created with the aid of pencil and paper or of high-speed computers, that represents a simplified version of reality and allows scientists to attempt to isolate and to understand the most important processes occurring in a specific situation.

modified Newtonian dynamics (MOND): A variant theory of gravity proposed by the Israeli physicist Mordehai Milgrom.

molecule: A stable grouping of two or more *atoms*.

mutation: A change in an organism's *DNA* that can be inherited by descendants of that organism.

natural selection: Differential success in reproduction among organisms of the same species, the driving force behind the *evolution* of life on Earth.

nebula: A diffuse mass of gas and dust, usually lit from within by young, highly luminous stars that have recently formed from this material.

neutrino: An *elementary particle* with no *electric charge* and a mass much smaller than an *electron*'s mass, characteristically produced or absorbed in reactions among elementary particles governed by *weak forces*.

neutron: An *elementary particle* with no *electric charge*; one of the two basic components of an atomic *nucleus*.

neutron star: The tiny remnants (less than twenty miles in diameter) of the core of a *supernova* explosion, composed almost entirely of *neutrons* and so dense that its matter effectively crams two thousand ocean liners into each cubic inch of space.

nitrogen: The element made up of *atoms* whose *nuclei* each have seven *protons*, and whose different *isotopes* have nuclei with six, seven, eight, nine, or ten neutrons. Most nitrogen nuclei have seven neutrons.

nuclear fusion: The joining of two *nuclei* under the influence of *strong forces*, which occurs only if the nuclei approach one another at a distance approximately the size of a proton (10^{-13} centimeter).

nucleic acid: Either *DNA* or *RNA*.

nucleotide: One of the cross-linking molecules in *DNA* and *RNA*. In DNA, the four nucleotides are adenine, cytosine, guanine, and thymine; in RNA, uracil plays the role that thymine does in DNA.

nucleus (pl. **nuclei**): (1) the central region of an *atom*, composed of one or more *protons* and zero or more *neutrons*. (2) The region within a *eukaryotic* cell that contains the cell's genetic material in the form of chromosomes. (3) The central region of a *galaxy*.

Oort cloud: The billions or trillions of *comets* that orbit the Sun, which formed first as the *protosun* began to contract, almost all of which move in orbits thousands or even tens of thousands of times larger than Earth's orbit.

organic: Referring to chemical compounds with *carbon atoms* as an important structural element; carbon-based molecules. Also, having properties associated with life.

organism: An object endowed with the property of being alive.

oxidation: Combination with *oxygen atoms*, typified by the rusting of metals upon exposure to oxygen in Earth's atmosphere.

oxygen: The element whose *nuclei* each have eight *protons*, and whose different *isotopes* each have seven, eight, nine, ten, eleven, or twelve neutrons in each nucleus. Most oxygen nuclei have eight neutrons to accompany their eight protons.

ozone (O_3): Molecules made of three *oxygen atoms*, which, at high altitudes in the Earth's atmosphere, shield Earth's surface against *ultraviolet* radiation.

panspermia: The hypothesis that life from one locale can be transferred to another, e.g., from planet to planet within the solar system; also called cosmic seeding.

photon: An elementary particle with no mass and no *electric charge*, capable of carrying *energy*. Streams of photons form *electromagnetic radiation* and travel through space at the speed of light, 299,792 kilometers per second.

photosynthesis: The use of *energy* in the form of *visible light* or *ultraviolet photons* to produce *carbohydrate* molecules from *carbon dioxide* and water. In some organisms, hydrogen sulfide (H_2S) plays the same role that water (H_2O) does in most photosynthesis on Earth.

planet: An object in orbit around another star that is not another star and has a size at least as large as Pluto, which ranks either as the Sun's smallest planet or as a *Kuiper Belt* object too small to be a planet.

planetesimal: An object much smaller than a planet, capable of building planets through numerous mutual collisions.

plate tectonics: Slow motions of plates of the crust of Earth and similar planets.

primitive atmosphere: The original atmosphere of a planet.

prokaryote: A member of one of the three domains of life, consisting of single-celled organisms in which the genetic material does not reside within a well-defined *nucleus* of the cell.

protein: A long-chain *molecule* made of one or more chains of *amino acids*.

proton: An *elementary particle* with one unit of positive *electric charge* found in the *nucleus* of every *atom*. The number of protons in an atom's *nucleus* defines the elemental identity of that atom. For example, the *element* that has one proton is *hydrogen*, the one with two protons is *helium*, and the element with ninety-two protons is uranium.

proton-proton cycle: The chain of three *nuclear fusion* reactions by which most stars fuse *protons* into *helium* nuclei and convert *energy* of mass into *kinetic energy*.

protoplanet: A planet during its later stages of formation.

protoplanetary disk: The disk of gas and dust that surrounds a *star* as it forms, from and within which individual planets may form.

protostar: A *star* in formation, contracting from a much larger cloud of gas and dust as the result of its self-gravitation.

pulsar: An object that emits regularly spaced pulses of radio *photons* (and often of higher-energy photons as well) as the result of the rapid rotation of a *neutron star*, which produces *radiation* as charged particles accelerate in the intense magnetic field associated with the neutron star.

quantum mechanics: The description of particles' behavior at the smallest scales of size, hence of the structure of *atoms* and their interaction with other atoms and *photons*, as well as the behavior of atomic *nuclei*.

quasar (quasi-stellar radio source): An object almost starlike in appearance, but whose *spectrum* show a large *red shift*, as a result of the object's immense distance from the *Milky Way*.

radiation: Short for *electromagnetic radiation*. In this nuclear age, the term has also come to mean any particle or form of light that is bad for your health.

radio: *Photons* with the longest *wavelengths* and lowest *frequencies*.

radioactive decay: The process by which certain types of atomic *nuclei* spontaneously transform themselves into other types.

red-giant star: A *star* that has evolved through its main sequence phase and has begun to contract its core and expand its outer layers. The contraction induces a greater rate of *nuclear* *fusion*, raises the star's *luminosity*, and deposits energy in the outer layers, thereby forcing the star to grow larger.

red shift: A shift to lower *frequencies* and longer *wavelengths* in the *spectrum* of an object, typically caused by the *Doppler effect*.

relativity: The general term used to describe Einstein's *special theory of relativity* and *general theory of relativity*.

replication: The process by which a "parent" *DNA* molecule divides into two single strands, each of which forms a "daughter" molecule identical to the parent.

resolution: The ability of a light-collecting device such as a camera, telescope, or microscope to capture detail. Resolution is always improved with larger lenses or mirrors, but this improvement may be negated by atmospheric blurring.

revolution: Motion around another object; for example, Earth revolves around the Sun. Revolution is often confused with *rotation*.

RNA (ribonucleic acid): A large, complex molecule, made of the same types of molecules that constitute *DNA*, which performs various important functions within living cells, including carrying the genetic messages embodied in DNA to the locations where *proteins* are assembled.

rotation: The spinning of an object on its own axis. For example, Earth rotates once every 23 hours and 56 minutes.

runaway greenhouse effect: A *greenhouse effect* that grows stronger as the heating of a planet's surface increases the rate of liquid evaporation, which in turn increases the greenhouse effect.

satellite: A relatively small object that orbits a much larger and more massive one; more precisely, both objects orbit their common center of mass, in orbits whose sizes are inversely proportional to the objects' masses.

self-gravitation: The *gravitational forces* that each part of an object exert on all the other parts.

SETI: The search for extraterrestrial intelligence.

shooting star: A popular name for a *meteor*.

skepticism: A questioning or doubting state of mind, which lies at the root of scientific inquiry into the cosmos.

Small Magellanic Cloud: The smaller of the two *irregular galaxies* that are satellites of our *Milky Way*.

solar system: The Sun plus the objects that orbit it, including *planets*, their *satellites*, *asteroids*, *meteoroids*, *comets*, and interplanetary dust.

solar wind: Particles ejected from the Sun, mostly *protons* and *electrons*, which emerge continuously from the Sun's outermost layers, but do so in especially large numbers at the time of an outburst called a solar flare.

solvent: A liquid capable of dissolving another substance; a liquid within which *atoms* and *molecules* can float and interact.

space-time: The mathematical combination of space and time that treats time as a coordinate with all the rights and privileges accorded space. It has been shown through the *special theory of relativity* that nature is most accurately described using a space-time formalism. It simply requires that all events are specified with space *and* time coordinates. The appropriate mathematics does not concern itself with the difference.

special theory of relativity: First proposed in 1905 by Albert Einstein, this provides a renewed understanding of space, time, and motion. The theory is based on two "Principles of Relativity": (1) the speed of light is constant for everyone no matter how you choose to measure it; and (2) the laws of physics are the same in every frame of reference that is either stationary or moving with constant velocity. The theory was later extended to include accelerating frames of reference in the *general theory of relativity*. It turns out that the two Principles of Relativity that Einstein *assumed* have been shown to be valid in every experiment ever performed. Einstein extended the relativity principles to their logical conclusions and predicted an array of unusual concepts, including:

- There is no such thing as absolute simultaneous events. What is simultaneous for one observer may have been separated in time for another observer.
- The faster you travel, the slower your time progresses relative to someone observing you.
- The faster you travel, the more massive you become, so the engines of your spaceship are less and less effective in increasing your speed.
- The faster you travel, the shorter your spaceship becomes—everything gets shorter in the direction of motion.
- At the speed of light, time stops, you have zero length, and your mass is infinite. Upon realizing the absurdity of this limiting case, Einstein concluded that you cannot reach the speed of light.

Experiments invented to test Einstein's theories have verified all of these predictions precisely. An excellent example is provided by particles that have decay "half-lives." After a predictable time, half are expected to decay into another particle. When these particles are sent to speeds near the speed of light (in particle accelerators), the half-life increases in the exact amount predicted by Einstein. They also get harder to accelerate, which implies that their effective mass has increased.

species: A particular type of organism, whose members possess similar anatomical characteristics and can interbreed.

spectrum (pl: **spectra**): The distribution of *photons* by *frequency* or *wavelength*, often shown as a graph that presents the number of photons at each specific frequency or wavelength.

sphere: The only solid shape for which every point on the surface has the same distance from the center.

spiral arms: The spiral features seen within the disk of a *spiral galaxy*, outlined by the youngest, hottest, most luminous stars and by giant clouds of gas and dust within which such stars have recently formed.

spiral galaxy: A *galaxy* characterized by a highly flattened disk of stars, gas, and dust, distinguished by *spiral arms* within the disk.

star: A mass of gas held together by its *self-gravitation*, at the center of which *nuclear fusion* reactions turn energy of mass into *kinetic energy* that heats the entire star, causing its surface to glow.

star cluster: A group of stars born at the same time and place, capable of enduring as a group for billions of years because of the stars' mutual gravitational attraction.

strong forces: One of the four basic types of *forces*, always attractive, that act between nucleons (*protons* and *neutrons*) to bind them together in atomic nuclei, but only if they approach one another within distances comparable to 10^{-13} cm.

sublimation: The transition from the solid to the gaseous state, or from gas to solid, without a passage through the liquid state.

submillimeter: *Electromagnetic radiation* with *frequencies* and *wavelengths* between those of *radio* and *infrared*.

supermassive black hole: A *black hole* with more than a few hundred times the mass of the Sun.

supernova (pl: **supernovae**): A *star* that explodes at the end of its *nuclear-fusing* lifetime, attaining such an enormous *luminosity* for a few weeks that it can almost equal the energy output of an entire *galaxy*. Supernovae produce and distribute *elements* heavier than *hydrogen* and *helium* throughout interstellar space.

telescope (gamma, X ray, ultraviolet, optical (visible), infrared, microwave, radio): Astronomers have designed special telescopes and detectors for each part of the *spectrum*. Some parts of this spectrum do not reach Earth's surface. To see the *gamma rays*, *X rays*, *ultraviolet*, and *infrared* that is emitted by many cosmic objects, these telescopes must be lifted into orbit above the absorbing layers of Earth's atmosphere. The telescopes are of different designs but they do share three basic principles: (1) They collect *photons*. (2) They focus photons. And (3) they record the photons with some sort of detector.

temperature: The measure of the average *kinetic energy* of random motion within a group of particles. On the *absolute* or *Kelvin temperature scale*, the temperature of a gas is directly proportional to the average kinetic energy of the particles in the gas.

thermal energy: The energy contained in an object (solid, liquid, or gaseous) by virtue of its atomic or molecular vibrations. The average *kinetic energy* of these vibrations is the official definition of temperature.

thermonuclear: Any process that pertains to the behavior of the atomic *nucleus* in the presence of high temperatures.

thermonuclear fusion: Another name for *nuclear fusion*, sometimes simply referred to as fusion.

thermophile: An organism that thrives at high temperatures, close to the boiling point of water.

tides: Bulges produced in a deformable object by the *gravitational force* from a nearby object, which arise from the fact that the nearby object exerts different amounts of force on different parts of the deformable object, since those parts have different distances from it.

UFOs (unidentified flying objects): Objects seen in the skies of Earth for which a natural explanation cannot be easily assigned, revealing either a profound ignorance within the scientific community or a profound ignorance within the observer.

ultraviolet radiation: *Photons* with *frequencies* and *wavelengths* between those of *visible light* and *X rays*.

universe: Usually taken to mean everything that exists, though in modern theories what we call the universe may prove to be only one part of a much larger "metaverse" or "multiverse."

virus: A complex of *nucleic acids* and *protein* molecules that can reproduce itself only within a "host" cell of another organism.

visible light: *Photons* whose *frequencies* and *wavelengths* correspond to those detected by human eyes, intermediate between those of *infrared* and *ultraviolet* radiation.

Voyager **spacecraft**: The two NASA spacecraft, named *Voyager 1* and *Voyager 2*, that were launched from Earth in 1978 and passed by Jupiter and Saturn a few years later; *Voyager 2* went on to encounter Uranus in 1986 and Neptune in 1989.

wavelength: The distance between successive wave crests or wave troughs; for *photons*, the distance that a photon travels while it oscillates once.

weak forces: One of the four basic types of *forces*, acting only among elementary particles at distances of about 10^{-13} cm or less, and responsible for the decay of certain elementary particles into other types. Recent investigations have shown that weak forces and *electromagnetic forces* are different aspects of a single *electroweak force*.

white dwarf: The core of a star that has fused *helium* into *carbon nuclei*, and therefore consists of carbon nuclei plus *electrons*, squeezed to a small diameter (about the size of Earth) and a high density (about 1 million times the density of water).

WMAP (Wilkinson Microwave Anisotropy Probe) satellite: The satellite launched in 2001 to study the *cosmic background radiation* in much greater detail than the *COBE satellite* could achieve.

X rays: *Photons* with *frequencies* greater than those of *ultraviolet* but less than those of *gamma rays*.

Further Reading

Adams, Fred, and Greg Laughlin. *The Five Ages of the Universe: Inside the Physics of Eternity*. New York: Free Press, 1999.

Barrow, John. *The Constants of Nature: From Alpha to Omega—The Numbers That Encode the Deepest Secrets of the Universe*. New York: Knopf, 2003.

———. *The Book of Nothing: Vacuums, Voids, and the Latest Ideas About the Origins of the Universe*. New York: Pantheon Books, 2001.

Barrow, John, and Frank Tipler. *The Anthropic Cosmological Principle*. Oxford: Oxford University Press, 1986.

Bryson, Bill. *A Short History of Nearly Everything*. New York: Broadway Books, 2003.

Danielson, Dennis Richard. *The Book of the Cosmos*. Cambridge, MA: Perseus, 2001.

Goldsmith, Donald. *Connecting with the Cosmos: Nine Ways to Experience the Majesty and Mystery of the Universe*. Naperville, IL: Sourcebooks, 2002.

———. *The Hunt for Life on Mars*. New York: Dutton, 1997.

———. *Nemesis: The Death-Star and Other Theories of Mass Extinction*. New York: Walker Books, 1985.

———. *Worlds Unnumbered: The Search for Extrasolar Planets.* Sausalito, CA: University Science Books, 1997.

———. *The Runaway Universe: The Race to Find the Future of the Cosmos.* Cambridge, MA: Perseus, 2000.

Gott, J. Richard. *Time Travel in Einstein's Universe: The Physical Possibilities of Travel Through Time.* Boston: Houghton Mifflin, 2001.

Greene, Brian. *The Elegant Universe.* New York: W. W. Norton & Co., 2000.

———. *The Fabric of the Cosmos: Space, Time, and the Texture of Reality.* New York: Knopf, 2003.

Grinspoon, David. *Lonely Planets: The Natural Philosophy of Alien Life.* New York: HarperCollins, 2003.

Guth, Alan. *The Inflationary Universe.* Cambridge, MA: Perseus, 1997.

Haack, Susan. *Defending Science—Within Reason.* Amherst, NY: Prometheus, 2003.

Harrison, Edward. *Cosmology: The Science of the Universe*, 2nd ed. Cambridge: Cambridge University Press, 1999.

Kirshner, Robert. *The Extravagant Universe: Exploding Stars, Dark Energy, and the Accelerating Cosmos.* Princeton, NJ: Princeton University Press, 2002.

Knoll, Andrew. *Life on a Young Planet: The First Three Billion Years of Evolution on Earth.* Princeton, NJ: Princeton University Press, 2003.

Lemonick, Michael. *Echo of the Big Bang.* Princeton, NJ: Princeton University Press, 2003.

Rees, Martin. *Before the Beginning: Our Universe and Others.* Cambridge, MA: Perseus, 1997.

———. *Just Six Numbers: The Deep Forces That Shape the Universe.* New York: Basic Books, 1999.

———. *Our Cosmic Habitat.* New York: Orion, 2002.

Seife, Charles. *Alpha and Omega: The Search for the Beginning and End of the Universe.* New York: Viking, 2003.

Tyson, Neil deGrasse. *Just Visiting This Planet: Merlin Answers More Questions About Everything Under the Sun, Moon and Stars.* New York: Main Street Books, 1998.

———. *Merlin's Tour of the Universe: A Skywatcher's Guide to Every-*

thing from Mars and Quasars to Comets, Planets, Blue Moons and Werewolves. New York: Main Street Books, 1997.

—————. *The Sky Is Not the Limit: Adventures of an Urban Astrophysicist.* New York: Doubleday & Co., 2000.

—————. *Universe Down to Earth.* New York: Columbia University Press, 1994.

—————, Robert Irion, and Charles Tsun-Chu Liu. *One Universe: At Home in the Cosmos.* Washington, DC: Joseph Henry Press, 2000.

Image Credits

Abbreviations
AURA: Association for University Research in Astronomy
CFHT: Canada, France, Hawaii Telescope
ESA: European Space Agency
ESO: European Southern Observatory
NASA: National Aeronautics and Space Administration
NOAO: National Optical Astronomical Observatories
NSF: National Science Foundation
USNO: United States Naval Observatory

1. WMAP Science Team, NASA
2. S. Beckwith and the Hubble Ultra Deep Field Working Group, ESA, NASA
3. Andrew Fruchter et al., NASA
4. N. Benitez, T. Broadhurst, H. Ford, M. Clampin, G. Hartig, and G. Illingworth, ESA, NASA
5. A. Siemiginowska, J. Bechtold, et al., NASA
6. O. Lopez-Cruz et al., AURA, NOAO, NSF
7. Jean-Charles Cuillandre, CFHT
8. Arne Henden, USNO

9. European Southern Observatory

10. Hubble Heritage Team, A. Riess, NASA

11. High-Z Supernova Search Team, NASA

12. Diane Zeiders and Adam Block, NOAO, AURA, NSF

13. P. Anders et al., ESA, NASA

14. Robert Gendler; www.robertgendlerastropics.com

15. Hubble Heritage Team, NASA

16. AURA/NOAO/NSF

17. M. Heydari-Malayeri (Paris Observatory) et al., ESA, NASA

18., 19. Atlas Image obtained as part of the Two Micron All Sky Survey, a joint project of the UMass and the IPAC/Caltech, funded by the NASA and the NSF.

20. Jean-Charles Cuillandre, CFHT

21. Jean-Charles Cuillandre, CFHT

22. J. Hester (Arizona State Univ.) et al., NASA

23. H. Bond and R. Ciardullo, NASA

24. Andrew Fruchter (Space Telescope Science Institute) et al., NASA

25. Jean-Charles Cuillandre, CFHT

26. Rick Scott; members.cox.net/rmscott

27. R. G. French, J. Cuzzi, L. Dones, and J. Lissauer, Hubble Heritage Team, NASA

28. (a) *Voyager 2*, NASA; (b) Athena Coustenis et al., CFHT

29. *Cassini* Imaging Team, NASA

30. (a) and (b) *Galileo* Project, NASA

31. *Magellan* Project, Jet Propulsion Laboratory, NASA

32. Buzz Aldrin, NASA

33. Juan Carlos Casado; www. skylook.net

34. J. Bell, M. Wolff, et al., NASA

35. *Spirit* rover, NASA/Jet Propulsion Laboratory/Cornell

36. *Spirit* rover, NASA/Jet Propulsion Laboratory/Cornell

37. Sandra Haller, Unicorn Projects, Inc.

38. Don Davis, NASA

39. Neil deGrasse Tyson, American Museum of Natural History

40. Sandra Haller, Unicorn Projects, Inc.

Index